JOURNAL OF APPLIED LOGICS - IFCOLOG
JOURNAL OF LOGICS AND THEIR APPLICATIONS

Volume 9, Number 2

April 2022

Disclaimer

Statements of fact and opinion in the articles in Journal of Applied Logics - IfCoLog Journal of Logics and their Applications (JALs-FLAP) are those of the respective authors and contributors and not of the JALs-FLAP. Neither College Publications nor the JALs-FLAP make any representation, express or implied, in respect of the accuracy of the material in this journal and cannot accept any legal responsibility or liability for any errors or omissions that may be made. The reader should make his/her own evaluation as to the appropriateness or otherwise of any experimental technique described.

ISBN 978-1-84890-397-5
ISSN (E) 2631-9829
ISSN (P) 2631-9810

College Publications
Scientific Director: Dov Gabbay
Managing Director: Jane Spurr

http://www.collegepublications.co.uk

EDITORIAL BOARD

SCOPE AND SUBMISSIONS

This journal considers submission in all areas of pure and applied logic, including:

pure logical systems
proof theory
constructive logic
categorical logic
modal and temporal logic
model theory
recursion theory
type theory
nominal theory
nonclassical logics
nonmonotonic logic
numerical and uncertainty reasoning
logic and AI
foundations of logic programming
belief change/revision
systems of knowledge and belief
logics and semantics of programming
specification and verification
agent theory
databases

dynamic logic
quantum logic
algebraic logic
logic and cognition
probabilistic logic
logic and networks
neuro-logical systems
complexity
argumentation theory
logic and computation
logic and language
logic engineering
knowledge-based systems
automated reasoning
knowledge representation
logic in hardware and VLSI
natural language
concurrent computation
planning

This journal will also consider papers on the application of logic in other subject areas: philosophy, cognitive science, physics etc. provided they have some formal content.

Submissions should be sent to Jane Spurr (jane@janespurr.net) as a pdf file, preferably compiled in LaTeX using the IFCoLog class file.

CONTENTS

ARTICLES

Causality and Responsibility in the Context of Multiple Agents

Robert Demolombe
No affiliation
robert.demolombe@orange.fr

Abstract

We present a definition of causality for cases where a state of affairs has been caused by a group of agents acting together. We use modal logic to formalize this definition and then we prove that the result of the actions was caused precisely by the agents of the group. In this paper we show another definition of causality in the case where the situation was caused independently by actions performed by several subsets of agents. In addition, we systematically analyze the notion of the level of responsibility of a group agent when the actions performed by the group have violated a norm. In this context, it is assumed that the level of responsibility relies on what the agents have done and also on what they believe about what has been done by each agent. Finally, we apply our formal definitions of causality on a simple case study in order to of allowing an intuitive understanding.

Keywords. Causality, responsibility, multi-agent systems, modal logic.

1 Introduction

In order to assign responsibility to agents, we need to know by who damages are caused. Although responsibility does not have to be identified with causation, there is a close link between norms and causation (see [26, 22, 16, 23, 7, 10, 8, 12, 14]) and the notion of causation plays a crucial role in attributing responsibility. In the case where several agents act together, or when they interact, it is very difficult to know who caused this or that state of affairs.

Consider, for example, the following academic example proposed by Lindahl in [18] and at a latter date by Broesen in [5]. Suppose that in order to kill a person, one only has to make him drink 4 grams of poison, and suppose that agents i and j have simultaneously put 2 grams of poison in the glass of agent l. After drinking what is in their glass, agent l dies. In this case, neither i nor j individually caused the 4 grams of poison to be in the glass, but this was achieved by their joint actions.

In this situation, both i and j have violated a norm and may be responsible for that violation[1]. If they are responsible, they may be subject to a penalty. However, the level of responsibility (and the level of penalty) may depend on several notions. In this paper, we have assumed, for the sake of simplicity, that this level depends only on what agents think about the situation. For instance, the penalty level of agent i may be lower if i does not believe he put 2 grams of poison in the glass, or if he believes that agent j did not put 2 grams of poison in the glass.

Now suppose that instead of acting simultaneously, the first agent i puts 2 grams of poison in the glass, and then in a second step j adds another 2 grams of poison. Is this situation the same as the previous one? Obviously not, since i is not responsible for the fact that j added 2 grams of poison, and only j is responsible for the fact that there are 4 grams of poison in the glass.

Another significant situation is where both i and j put 4 grams of poison in the glass at the same time. According to most definitions of causation, an agent has caused a state of affairs if, in a situation where it did not act, that state of affairs would not have occurred. In this example, if i had not put poison in the glass, would there not have been 4 grams of poison in the glass? The answer is no, because in the counterfactual situation where i does not act, j always acts. The same type of argument could be used to conclude that j did not cause the presence of 4 grams of poison in the glass, and the final conclusion would be that neither i nor j is responsible for this state of affairs.

It is obvious that there is something wrong with this kind of reasoning. However, it is not insignificant to specify where exactly the error lies. This is why we must reconsider the formal definition of causality when several agents act "together". This is the purpose of the work we present in this paper.

We first define the logical framework that is used for the formal definitions and properties of the operators defining causality. Then, we propose a systematic analysis of the levels of responsibility when a group of agents has violated a norm. Finally, we reformulate in this logical context the case study that we presented earlier in the introduction.

2 Logical framework

Our logical framework is inspired by von Wright [26], Pörn [22] and Hilpinen [13]. It could also be seen as an extension to multiple agents of the framework proposed by Demolombe in [8]. Logic will only be defined in semantics since the main purpose

---

[1] It may be that, according to the norms, the responsible agents are not the ones who caused the violation (for example, if i is under 18 years of age).

is to clarify the meaning of concepts (interesting insights of the different approaches to action logic can be found in [13], [2] and [24]).

The main idea is that in the logical framework, the semantics of actions is defined by the effects caused by these actions, and also by the names of actions whose meaning is defined outside the logical framework.

From a theoretical point of view, we accept that the meaning of an action can be entirely defined by the set of all the effects that are caused by this action. For example, if we ask: *What is the meaning of the action that is called "closing the door"?* the answer is *It is an action that causes the door to be closed.* Nevertheless, there are many different ways of closing the door, and when we don't need to distinguish between these different ways, it is more convenient to use the term "closing the door" as the name of a type of action that can be performed in different ways. For instance, it can be closed slowly or quickly.

Obviously, we also need to identify the agent that performs an action. The formal consequence of this approach is that action operators are identified by the effects they cause, as well as by the name of an agent and a type of action (when there is no ambiguity, we use the term "action" instead of "type of action"). A pair <agent, action type> is called a "**act**".

The language of the logic is defined using the following notation:

$ATOM$: set of atomic propositions denoted by p, q, r, ...

$AGENT$: set of agents denoted by i, j, k, l, ...

$ACTION$: set of action types [2] denoted by α, β, γ, δ, ...

ACT: set of acts denoted by pairs of the form $i : \alpha$, where i is in $AGENT$ and α is in $ACTION$.

$SACT$: set of sets of acts denoted by Act_1, Act_2, Act_3, ...

$SSACT$: set of sets of sets of acts denoted by Act_1^*, Act_2^*, Act_3^*, ...

An example of an element of $SACT$ is $\{i : \alpha, i : \beta, j : \gamma\}$. An example of an element of $SSACT$ is $\{\{i : \alpha, i : \beta, j : \gamma\}, \{k : \alpha, k : \delta\}\}$.

The language L is the set of formulas defined by the following BNF:

$$\phi ::= p \mid \neg\phi \mid \phi \vee \phi \mid Does_{Act[All]}\phi \mid Done_{Act[All]}\phi \mid JBri_{Act[All]}\phi \mid$$
$$RJBri_{Act,Act'[All]}\phi \mid SJBri_{Act^*[All]}\phi$$

---

[2]We may use the term "action" instead of "action type" for the sake of simplicity.

where p ranges over $ATOM$, Act and Act' range over $SACT$ and Act^* ranges over $SSACT$. We assume that Act, Act' and Act^* are non-empty sets. In the modal operators the index $[All]$ designates the set of all the acts performed in a given world. In operators such as $Does_{Act[All]}$ or $JBri_{Act[All]}$, Act designates a subset of the acts performed in All that have a specific effect.

The intuitive reading of the modal operators is as follows :

- $Does_{Act[All]}\phi$: the agents in Act are going to do the acts in Act and after their performance ϕ will be true.

- $Done_{Act[All]}\phi$: the agents in Act have just done the acts in Act and before their performance ϕ was true.

- $JBri_{Act[All]}\phi$: the agents in Act are going to bring it about that ϕ by doing the set of acts Act (the "J" in $JBri$ stands for "joint" acts).

- $RJBri_{Act,Act'[All]}\phi$: the agents in Act are going to bring it about that ϕ by doing the set of acts Act while the acts in Act' are not performed (the "R" in $RJBri$ stands for "restricted" joint acts). We assume that $Act \cap Act' = \emptyset$.

- $SJBri_{Act^*[All]}\phi$: every member Act of Act^* is going to bring it about that ϕ, even if the other members of Act^* are not performed (the "SJ" in $SJBri$ stands for "set of joint" acts).

It is assumed that in these definitions Act and Act' are included in All.

We shall use the notation $Does_{Act[All]}$ (respectively $Done_{Act[All]}$) as an abbreviation of $Does_{Act[All]}(true)$ (respectively $Done_{Act[All]}(true)$). The intuitive meaning of $Does_{Act[All]}$ (respectively $Done_{Act[All]}$) is that agents in Act are going to do (respectively, have just done) the acts in Act.

Definition 1. *A frame F is a tuple $F = < W, R^*_{Act}, CR^*_{Act,Act'} >$, where W is a non-empty set of worlds, Act and Act' are subsets of ACT, R^*_{Act} is a set of binary relations defined on $W \times W$ and $CR^*_{Act,Act'}$ is a set of ternary relations defined on $W \times W \times W$. In a frame we have the property: $(R_{Act}(w, w')$ and $R_{Act'}(w, w'))$ iff $R_{Act \cup Act'}(w, w')$.*

*A **model** M is a tuple $M = < F, v >$, where F is a frame and v is a function which assigns to each atomic proposition a subset of W.*

Relations have the following intuitive meaning:

- $R_{Act}(w, w')$ iff the performance of all acts in Act started in w and ended in w'.

- $R_{Act[All]}(w, w')$ iff the performance of all the acts in All started in w and ended in w', and Act is a subset of All.

- $R_{Act,Act'}(w, w', w'')$ iff $R_{Act}(w, w')$ and the only difference between w'' and w' is that in w'' none of the acts in Act' has been performed and the acts in $Act - Act'$ have been performed in w'' in the same way as in w' (*ceteris paribus* condition[3]).

In the following, if $R_{Act,Act'}(w, w', w'')$ holds we say, in short, that w'' *is a counterfactual world of w' with respect to Act'.*

Definition 2. *The fact that a formula ϕ is **true** in a world w of a model M is denoted by $M, w \models \phi$. The fact that ϕ is a **valid** formula, that is, ϕ is true in every world of every model, is denoted by $\models \phi$.*

The truth conditions for atomic propositions and logical connectives are defined as usual.

Definition 3. *We have the following truth conditions for the action operators of the kind Does and Done.*
  $M, w \models Does_{Act[All]}\phi$ *iff*
there exists a world w' such that $R_{Act[All]}(w, w')$ and for all w' $(R_{Act[All]}(w, w')$ $\Rightarrow$ $M, w' \models \phi)$.
  $M, w \models Done_{Act[All]}\phi$ *iff*
there exists w' such that $(R_{Act[All]}(w', w)$ and $M, w' \models \phi)$.

In the following it will be assumed that for every w, w', w'', Act_1 and Act_2: $R_{Act_1}(w', w)$ and $R_{Act_2}(w'', w)$ entail $w' = w''$. Intuitively, this means that the past is unique.

Properties.
We have the following properties.
  (D1) For every Act, Act' and All in ACT we have:
$\models Done_{Act[All]}\phi \wedge Done_{Act'[All]}\psi) \to Done_{Act \cup Act'[All]}(\phi \wedge \psi)$
  (D2) For every Act and All in ACT, if $Act \subseteq All$ we have:
$\models Done_{Act[All]}\phi \to Done_{All[All]}\phi$
  (D3) For every $i : \alpha$ in Act we have:
$\models Done_{Act[All]}\phi \to Done_{i:\alpha}\phi$

---

[3]The precise definition of the *ceteris paribus* condition raises very challenging problems which are ignored in this paper (see Lewis [17]).

Definition 4. *We have the following truth conditions for the operators of* **joint action** *of type* $JBri$.

$M, w \models JBri_{Act[All]}\phi$ *iff*

1. *for every* w' $(R_{All}(w, w') \Rightarrow M, w' \models \phi)$

2. *for every* $i : \alpha$ *in* Act, *there exist* w' *and* w'' *such that*
   $R_{All,\{i:\alpha\}}(w, w', w'')$ *and* $M, w'' \models \neg\phi$

3. *for every non empty set* Sub *such that* $Sub \subseteq All - Act$, *for every* w' *and* w''
   $R_{All,Sub}(w, w', w'') \Rightarrow M, w'' \models \phi$.

The intuitive meaning of condition 1) is that the set of acts in All is **sufficient** to ensure that we get ϕ. The intuitive meaning of condition 2) is that each act $i : \alpha$ in Act is **necessary** to obtain ϕ. That is, if $i : \alpha$ is the only act in All that is not performed, there is a world w'' where ϕ is not obtained.

The intuitive meaning of condition 3) is that even if the acts in Sub are not performed, the acts in Act are **sufficient** to obtain ϕ.

If Act contains only one act, we write: $JBri_{i:\alpha[All]}\phi \stackrel{\text{def}}{=} JBri_{\{i:\alpha\}[All]}\phi$.

Example 1. The general definitions in the Definition 4 are applied to the case study presented in the introduction. We use the following notations: $add(N)$: type of action "to put N grams of poison in the glass", $add(0)$: type of action that has no effect on the content of the glass, $pois(N)$ is an atomic proposition which means "there are N grams of poison in the glass".

It is assumed that we have: $All = \{i : add(1), j : add(3), k : add(0)\}$ and $Act = \{i : add(1), j : add(3)\}$.

It is assumed that we are in a world w of a model M where we have: $M, w \models pois(0)$. Then, we have: $R_{All}(w, w')$ and $M, w' \models pois(4)$.

If action $i : add(1)$ had not been realized we would be in a world w''_1 such that: $R_{All,i:add(1)}(w, w', w''_1)$ and $M, w''_1 \models pois(3)$.

If action $i : add(3)$ had not been realized we would be in a world w''_2 such that: $R_{All,i:add(3)}(w, w, w''_2)$ and $M, w''_2 \models pois(1)$.

If action $i : add(0)$ had not been realized we would be in a world w''_3 such that: $R_{All,i:add(0)}(w, w', w''_3)$ and $M, w''_3 \models pois(4)$.

Therefore, we have: $M, w \models JBri_{Act[All]}(pois(4))$.

It is assumed that $add(-N)$ is an action that neutralizes the effects of N grams of poison. If we add the action $add(-2)$ to the set of acts All, we have:

$R_{All}(w, w')$ implies $M, w' \models pois(2)$

Therefore, $M, w \models JBri_{Act[All]}(pois(4))$ is false because the condition 1 in Definition 4 is false. This case shows the role of condition 1.

Theorem 1. *If $Act' \subset Act$, we have:*

$$(NM1) \models JBri_{Act[All]}\phi \rightarrow \neg JBri_{Act'[All]}\phi.$$

If $Act \subset Act''$, we have:

$$(NM2) \models JBri_{Act[All]}\phi \rightarrow \neg JBri_{Act''[All]}\phi.$$

Proof: the proofs are similar as the proofs in [9].[4]

The intuitive reading of Theorem 1 is that if $JBri_{Act[All]}\phi$ holds, ϕ was caused by the performance of the acts in Act and by no other act.

Theorem 2. *The following statements hold:*

$$(CL) \models (JBri_{Act_1[All]}\phi \wedge JBri_{Act_2[All]}\psi) \rightarrow JBri_{Act_1 \cup Act_2[All]}(\phi \wedge \psi).$$
$$(CL') \models (JBri_{Act[All]}\phi \wedge JBri_{Act[All]}\psi) \rightarrow JBri_{Act[All]}(\phi \wedge \psi).$$

(NCL) *If we have $\models \phi \rightarrow \psi$ and $Act_2 \nsubseteq Act_1$ or $\models \psi \rightarrow \phi$ and $Act_1 \nsubseteq Act_2$,*

then we have $\models (JBri_{Act_1[All]}\phi \wedge JBri_{Act_2[All]}\psi) \rightarrow \bot$.

Proof: the proofs are similar as the proofs in [9].

It is worth noting that from (NCL) if $Act_1 \neq Act_2$ we have $\models (JBri_{Act_1[All]}\phi \wedge JBri_{Act_2[All]}\phi) \rightarrow \bot$. From (CL) we also have: $\models (JBri_{Act_1[All]}\phi \wedge JBri_{Act_2[All]}\phi) \rightarrow JBri_{Act_1 \cup Act_2[All]}\phi$. This property does not contradict the non monotonicity theorems (NM1) and (NM2). Indeed, if $Act_1 \neq Act_2$ the antecedent $JBri_{Act_1[All]}\phi \wedge JBri_{Act_2[All]}\phi$ of the implication is inconsistent, and if $Act_1 = Act_2$ we have: $Act_1 \cup Act_2 = Act_1$.

Theorem 3. *The following statements hold:*
$(\neg N) \not\models JBri_{Act[All]}(\top)$.
$(\neg DM)$ *If there is more than one act in* Act *and we have $\not\models \phi \rightarrow \psi$ and $\not\models \psi \rightarrow \phi$, then we have $\not\models JBri_{Act[All]}(\phi \wedge \psi) \rightarrow (JBri_{Act[All]}\phi \vee JBri_{Act[All]}\psi)$.*
$(\neg M)$ *If there is more than one act in* Act *and we have $\not\models \phi \rightarrow \psi$ and $\not\models \psi \rightarrow \phi$, then we have*
$\not\models JBri_{Act[All]}(\phi \wedge \psi) \rightarrow JBri_{Act[All]}\phi$ *and*
$\not\models JBri_{Act[All]}(\phi \wedge \psi) \rightarrow JBri_{Act[All]}\psi$.

Proof: see [9] for the proofs of $(\neg N)$ and $(\neg M)$. There is an error in the proof of $(\neg DM)$. See the Annex for the new proof.

---

[4]In Definition 4 the condition 3 is not the same as its definition in [9]. However it is similar, and the definition of condition 3 in [9] is a special case of the new definition. Then, the proofs can easily be translated and the results are the same.

It should be noted that ($\neg$DM) results false in the particular case where there is only one act in *Act*. Indeed, in this case, we have: $\models JBri_{Act[All]}(\phi \wedge \psi) \rightarrow (JBri_{Act[All]}\phi \vee JBri_{Act[All]}\psi)$.

Definition 5. *The following are the truth conditions for the operators of* **restricted joint action** *of the kind RJBri. It is assumed that in Act* there is more than one set of acts.*

$M, w \models RJBri_{Act,Act'[All]}\phi$ *iff*

1. *for all w' ($R_{All}(w, w') \Rightarrow M, w' \models \phi$)*

2. *for all $i : \alpha$ in Act, there exist w', w'' and w''' such that*
   ($R_{All}(w, w')$ and $R_{All,Act'}(w, w', w'')$ and $R_{(All-Act'),\{i:\alpha\}}(w, w'', w''')$ and $M, w''' \models \neg\phi$)

3. *for every non empty set Sub such that $Sub \subseteq All - Act$ and for every w' and w'' $R_{All,Sub}(w, w', w'') \Rightarrow M, w'' \models \phi$*

Condition 1) shows that acts in *All* are sufficient to obtain ϕ. The intuitive meaning of condition 2) is that w'' is a counterfactual world of w' when *Act'* is not performed, and w''' is a counterfactual world of w'' when $i : \alpha$ is not executed. Then, condition 2) shows that when *Act'* is not executed $i : \alpha$ is a necessary act to get ϕ. The meaning of condition 3) is that acts that are not in *Act* are not necessary to get ϕ.

From an intuitive point of view in Definition 5 the effects of the acts in *Act'* are ignored.

For the operator $SJBri_{Act^*[All]}$ it is assumed that the members Act_i of Act^* are all disjoint.

Definition 6. *We have the following truth conditions for the operators of* **set of joint actions** *of the kind SJBri.*

$M, w \models SJBri_{Act^*[All]}\phi$ *iff*

1. *for all w' ($R_{All}(w, w') \Rightarrow M, w' \models \phi$)*

2. *for every Act_i in Act*, for every non empty set Sub such that $Sub \subseteq All - Act_i$ and for every w' and w'' $R_{All,Sub}(w, w', w'') \Rightarrow M, w'' \models \phi$*

3. *for every Act_i in Act*:*
   for every $i : \alpha$ in Act_i there exist w', w'' and w''' such that
   ($R_{All}(w, w')$ and $R_{All,Act^*-Act_i}(w, w', w'')$ and
   $R_{All-(Act^*-Act_i),\{i:\alpha\}}(w, w'', w''')$ and $M, w''' \models \neg\phi$).

We have used, as an abuse of rating, the notation $Act^* - Act_i$ to represent the difference between the set of all the acts in the set of acts Act^* and the set of acts in Act.

In Definition 6, in a world w'' such that $R_{All,Act^*-Act_i}(w, w', w'')$ the only acts in Act^* which have been performed are the acts in Act_i and in a world w''' such that $R_{All-(Act^*-Act_i),\{i:\alpha\}}(w, w'', w''')$ the only act in Act_i which is not performed is $i : \alpha$.

Condition 1) is the same as for modality $JBri$.

In condition 2) the world w'' is a counterfactual world of w' with respect to acts in Act^* that are not in Act_i. That is, roughly speaking, in w'' Act_i is the only member of Act^* that is acting. The intuitive meaning of this condition is that each Act_i is sufficient to get ϕ. Observe that condition 1) is not redundant with condition 2). Indeed, in condition 2), the Act_i members of Act^* act independently, whereas condition 1) guarantees that they can act together.

In condition 3) the world w'' is a counterfactual world of w'' with respect to the acts of $Act^* - Act_i$, and w'' is a counterfactual world of w'' with respect to the act $i : \alpha$. That is to say that in w''', none of the acts in $Act^* - Act_i$ are executed and all acts are in Act_i but $i : \alpha$ are executed. The intuitive meaning of this condition is that all acts in Act_i are necessary to get ϕ in a context where members of Act^* that are different from Act_i do not act.

From conditions 2) and 3) it can be inferred that each Act_i in Act^* is a **sufficient** and **necessary** set of acts to bring it about that ϕ when the other members of Act^* are not acting.

Example 2. The general definitions in Definition 6 are applied to a similar example as Example 1.

We have the new notation: $poisgte(N)$ is a proposition that means that the number of grams of poison in the glass is greater or equal to N (abbreviated as "gte").

It is assumed that we have:
$All = \{i : add(1), j : add(3), i' : add(2), j' : add(2), k : add(0)\}$,
$Act_1 = \{i : add(1), j : add(3)\}$ and $Act_2 = \{i' : add(2), j' : add(2)\}$ and
$Act^* = \{Act_1, Act_2\}$

It is also assumed that we are in a world w of a model M where we have: $M, w \models pois(0)$.

Then we have: $R_{All}(w, w')$ and $M, w' \models poisgte(4)$.

We also have: $R_{All,Act^*-Act_1}(w, w'_1, w''_1)$ and $M, w''_1 \models poisgte(4)$

In w''_1 the unique set of acts in Act^* which is performed is Act_1.

In the same way we have: $R_{All,Act^*-Act_2}(w, w'_2, w''_2)$ and $M, w''_2 \models poisgte(4)$.

We have:

(1) $R_{All-(Act^*-Act_1),\{i:add(1)\}}(w, w''_1, u'''_1)$ and $M, u'''_1 \models pois(3)$

because in u'''_1 the acts that have been performed are the same as in w''_2 but the act

591

$i : add(1)$.

For similar reasons we have:

(2) $R_{All-(Act^*-Act_1),\{j:add(3)\}}(w, w_1'', u_2''')$ and $M, u_2''' \models pois(1)$ because $j : add(3)$ in Act_1 is not performed.

Properties (1) and (2) show that, when Act_2 is not performed, each act $i : add(1)$ and $j : add(3)$ is necessary to obtain $poisgte(4)$.

We also have:

(3) $R_{All-(Act^*-Act_2),\{i':add(2)\}}(w, w_1'', v_1''')$ and $M, v_1''' \models pois(2)$ because $i' : add(2)$ in Act_2 is not performed.

(4) $R_{All-(Act^*-Act_2),\{j':add(2)\}}(w, w_1'', v_2''')$ and $M, v_2''' \models pois(2)$ because $j' : add(2)$ in Act_2 is not performed.

Properties (3) and (4) show that, when Act_1 is not performed, each act $i' : add(2)$ and $j' : add(2)$ is necessary to obtain $poisgte(4)$.

Therefore we have: $M, w \models SJBri_{Act^*[All]}(poisgte(4))$.

Roughly speaking, the sets Act_1 and Act_2, acting alone, can cause $poisgte(4)$.

It is worth noting that we do not have $JBri_{Act_1[All]}pois(4)$ because the condition 2 in the definition 4 is not fulfilled . Indeed, even if $i : add(1)$ is not performed, we have $pois(7)$ due to the fact that Act_2 is performed in this situation.

This example shows that we need the modality $RJBri$. Indeed, we can check that we have $RJBri_{Act_1,Act_2[All]}pois(4)$ and $RJBri_{Act_2,Act_1[All]}pois(4)$. Intuitively, in the first case it is assumed that Act_2 is not performed and in the second Act_1 is not performed.

Theorem 4. *If Act_i is in Act^*, then*
$$\models SJBri_{Act^*[All]}\phi \rightarrow RJBri_{Act_i,(Act^*-Act_i)[All]}\phi.$$

Proof. This a trivial consequence of the truth conditions of the operator $SJBri_{Act^*[All]}$.

This theorem validates the idea that each set of acts in Act^* can cause ϕ independently of the other members of Act^*.

Definition 7. *The following modal operator is defined from the operators Done and JBri.*
$$JBro_{Act}\phi \stackrel{def}{=} Done_{All}(JBri_{Act[All]}\phi)$$

The intuitive meaning of the operator $JBro_{Act}\phi$ is that the set of agents in Act has brought it about that ϕ by doing exactly the set of acts Act.

Theorem 5. *The following property holds:*

$$(JT) \models JBro_{Act}\phi \rightarrow \phi$$

Proof: see [9].

3 Responsibility

This section discusses the levels of responsibility when a norm has been violated. Since responsibility depends on norms, it has no single definition. Thus, this work aims to define a general methodology that could be specified in different contexts.

At first step, we need to define the set of distinct situations that may occur. In a second step, we define an order, in terms of levels of responsibility, on this set of situations. First, we assume that responsibility is characterized by several general properties and shows that the number of distinct situations can be quite large.

Then, in order to make the problem easier to analyze, we restrict these properties to the acts that agents have performed and to what they believe about these acts. In this context we present a method to order the responsibility levels depending on the situations.

3.1 Set of all the situations in the case of a unique act

To limit the number of situations, we consider all the situations that can occur when a single act is performed by a i agent. This act is designated by: $i : a$.

Where a standard has been violated by the single act act, it is assumed that i's responsibility depends on the following properties: whether the act was performed, whether i has the capacity to perform that act, whether it intended to perform that act, and what the agent believes about those facts.

Notations [5]:
- $Bel_i(\phi)$: i believes ϕ,
- $Cap_i(i : a)$: i has the capability to perform $i : a$,
- $Int_i(i : a)$: i's intention is to perform $i : a$,
- $Int_i(\neg(i : a))$: i's intention is to do not perform $i : a$.

The set of all the situations is represented by the conjunction of the formulas:

(1) $(Done_{i:a} \vee \neg Done_{i:a})$
(2) $(Cap_i(i : a) \vee \neg Cap_i(i : a))$
(3) $(Int_i(i : a) \vee \neg Int_i(i : a))$
(4) $(Int_i(\neg(i : a)) \vee \neg Int_i(\neg(i : a)))$
(5) $(Bel_i(Done_{i:a}) \vee \neg Bel_i(Done_{i:a}))$
(6) $(Bel_i(\neg Done_{i:a}) \vee \neg Bel_i(\neg Done_{i:a}))$
(7) $(Bel_i(Cap_i(i : a)) \vee \neg Bel_i(Cap_i(i : a)))$
(8) $(Bel_i(\neg Cap_i(i : a)) \vee \neg Bel_i(\neg Cap_i(i : a)))$

---

[5] As a matter of simplification, when there is no risk of misinterpretation, formulas of the kind $Done_{Act[All]}\phi$ are abbreviated as $Done_{Act}\phi$.

(9) $(Bel_i(Int_i(i:a)) \lor \neg Bel_i(Int_i(i:a))$

(10) $(Bel_i(\neg Int_i(i:a)) \lor \neg Bel_i(\neg Int_i(i:a)))$

This formula characterizes 1.024 distinct situations.

It seems reasonable to accept the following assumptions :

(H1) $\neg(Int_i(\phi) \land Int_i(\neg(\phi)))$

(H2) $\neg(Bel_i(\phi) \land Bel_i(\neg\phi))$

(H3) $Int_i(\phi) \to Bel_i(Int_i(\phi))$

(H4) $\neg Int_i(\phi) \to Bel_i(\neg Int_i(\phi))$

(H5) $Done_{i:a} \to Cap_i(i:a)$

In addition it is assumed that the Bel_i operator is normal and closed under implication.

The meaning of (H1) and (H2) is that the agent i has consistent intentions and beliefs. The meaning of (H3) and (H4) is that the agent i is aware of its intentions. The meaning of (H5) is that if the agent i has performed an act, then he has the capacity to perform it.

Notations:

$Ign_i(\phi) \overset{\text{def}}{=} \neg Bel_i(\phi) \land \neg Bel_i(\neg\phi)$

$NoInt_i(\phi) \overset{\text{def}}{=} \neg Int_i(\phi) \land \neg Int_i(\neg\phi)$

The formulas (1) to (10) are logically equivalent to the conjunction of the formulas:

(1.2) $(Done_{i:a} \lor (\neg Done_{i:a} \land Cap_i(i:a)) \lor \neg Cap_i(i:a))$, from (1), (2) and (H5)

(3.4) $(Int_i(i:a) \lor NoInt_i(i:a) \lor Int_i(\neg i:a))$, from (3), (4) and (H1)

(5.6) $(Bel_i(Done_{i:a}) \lor Ign_i(Done_{i:a}) \lor Bel_i(\neg Done_{i:a}))$

(7.8) $(Bel_i(Cap_i(i:a)) \lor Ign_i(Cap_i(i:a)) \lor Bel_i(\neg Cap_i(i:a)))$, from (7), (8) and (H2)

(9.10) $(Bel_i(Int_i(i:a)) \lor Ign_i(Int_i(i:a)) \lor Bel_i(\neg Int_i(i:a)))$, from (9), (10) and (H2)

These formulas characterize 243 distinct situations. They are logically equivalent to the conjunction of the following ones:

(1.2) $(Done_{i:a} \lor (\neg Done_{i:a} \land Cap_i(i:a)) \lor \neg Cap_i(i:a))$

(3.4) $(Int_i(i:a) \lor NoInt_i(i:a) \lor Int_i(\neg i:a))$,

from (3.4), (9.10), (H3) and (H4)

(5.6.7.8) $(Bel_i(Done_{i:a}) \lor (Ign_i(Done_{i:a}) \land Bel_i(Cap_i(i:a)))\lor$
$(Ign_i(Done_{i:a}) \land Ign_i(Cap_i(i:a))) \lor Bel_i(\neg Cap_i(i:a))$

(9.10) $(Bel_i(Int_i(i:a)) \lor Ign_i(Int_i(i:a)) \lor Bel_i(\neg Int_i(i:a)))$

This formula characterizes 108 distinct situations. An example of such situations is represented by the formula:

(S) $\neg Cap_i(i:a)) \land Int_i(i:a) \land Bel_i(Done_{i:a})$

The intuitive meaning of (S) is that the agent i intended to do $i : a$ and believes he did, when he does not have the ability to do it.

In the context of our assumptions (S) is logically equivalent to:

(S') $\neg Cap_i(i : a)) \wedge \neg Done_{i:a} \wedge Int_i(i : a) \wedge Bel_i(Int_i(i : a)) \wedge Bel_i(Done_{i:a})$

The intuitive meaning of (S') is that the agent i intended to execute $i : a$, and he is aware of his intention, and he believes that he did $i : a$, whereas he did not execute $i : a$ because he does not have the ability to do so.

To simplify the analysis of responsibility, in what follows, the properties of capacity and intent are ignored.

3.2 Responsibility in the case of a set of acts

Here we analyse responsibility in contexts where is performed a set of acts Act. It is assumed that norms are of the form:

- it is obligatory that the set of agents in Act brings it about that ϕ by doing the set of acts in Act, or

- it is prohibited that the set of agents in Act brings it about that ϕ by doing the set of acts in Act.

Obligations are represented by formulas of the form: $Obg(JBri_{Act}(\phi))$, and prohibitions are represented by: $Proh(JBri_{Act}(\phi))$, where $Proh(JBri_{Act}(\phi))$ is a notation for: $Obg(\neg JBri_{Act}(\phi))$.

In the following it is accepted that lawyers assume that the performance of Act causes ϕ, and that $Obg(JBri_{Act}(\phi))$ is violated iff the acts in Act have not been performed, i.e. when we have: $\neg Done_{Act}$, and that $Proh(JBri_{Act}(\phi))$ is violated iff we have: $Done_{Act}$.

In order to set the level of responsibility of an agent i involved in Act in relation to the fact that he has done or he has not done an action of type a we must consider, on the one hand, the act $i : a$ and, on the other hand, the complementary set of acts: $Act - \{i : a\}$. Indeed, the level of responsibility of i [6] depends on what it Done and also on what other agents have done. More formally, we must consider the facts represented by: $Done_{i:a}$ and $Done_{Act-\{i:a\}}$.

Consequently, the norm $Obg(JBri_{Act}(\phi))$ is violated iff we have:
$\neg(Done_{i:a} \wedge Done_{Act-\{i:a\}})$, that is logically equivalent to:
$(\neg Done_{i:a} \wedge Done_{Act-\{i:a\}}) \vee (\neg Done_{i:a} \wedge \neg Done_{Act-\{i:a\}}) \vee$
$(Done_{i:a} \wedge \neg Done_{Act-\{i:a\}})$.

---

[6]This is only an example since the definition of responsibility depends on the norms of each institution.

The norm $Proh(JBri_{Act}(\phi))$ is violated iff all the acts in Act are performed, i.e. when we have: $Done_{i:a} \wedge Done_{Act-\{i:a\}}$.

It is assumed that agents know what the norms are and that they know that performing the set of acts Act causes ϕ (i.e. $JBri_{Act}(\phi)$).

The agent i's level of responsibility i can be assessed on the basis of what has been done (or not done) in a given situation, as well as the agent i's representation of that situation. For example, the agent i's level of responsibility may depend on whether $i : a$ has been performed and whether the agent i believes that $i : a$ has not been performed, or vice versa.

According to these assumptions, a complete representation of all the possible cases where we have to fix agent i's responsibility level is defined on the basis of the following information:

- act $i : a$ has been performed,

- the set of acts $Act - \{i : a\}$ has been performed,

- agent i believes that act $i : a$ has been performed,

- agent i believes that act $i : a$ has not been performed,

- agent i believes that the set of acts $Act - \{i : a\}$ has been performed

- agent i believes that the set of acts $Act - \{i : a\}$ has not been performed.

The set of all these situations is represented by the conjunction of the formulas:

(S1) $(Done_{i:a} \vee \neg Done_{i:a})$
(S2) $(Done_{Act-\{i:a\}} \vee \neg Done_{Act-\{i:a\}})$
(S3) $(Bel_i(Done_{i:a}) \vee \neg Bel_i(Done_{i:a}))$
(S4) $(Bel_i(\neg Done_{i:a}) \vee \neg Bel_i(\neg Done_{i:a}))$
(S5) $(Bel_i(Done_{Act-\{i:a\}}) \vee \neg Bel_i(Done_{Act-\{i:a\}}))$
(S6) $(Bel_i(\neg Done_{Act-\{i:a\}}) \vee \neg Bel_i(\neg Done_{Act-\{i:a\}}))$

This conjunction is logically equivalent to:

(S1) $(Done_{i:a} \vee \neg Done_{i:a})$
(S2) $(Done_{Act-\{i:a\}} \vee \neg Done_{Act-\{i:a\}})$
(S3.4) $(Bel_i(Done_{i:a}) \vee Ign_i(Done_{i:a}) \vee Bel_i(\neg Done_{i:a}))$, from (S3), (S4) and (H2)
(S5.6) $(Bel_i(Done_{Act-\{i:a\}}) \vee Ign_i(Done_{Act-\{i:a\}}) \vee Bel_i(\neg Done_{Act_{\{i:a\}}}))$, from (S5), (S6) and (H2)

In order to set the level of responsibility in each situation, we assume that this level depends on whether the representation of the world by the agent i is valid or not.

According to this general principle, we assume that the level is maximal if the agent has a true representation of the world, since in this case he is perfectly aware of what he has done. The level is minimum if he has a wrong representation of the world, since we can assume that he did not intend to do what he has done. When the agent is unaware of what he or she has done, the level of responsibility is between maximum and minimum, since it can be assumed that even if he or she did not intend to violate the standard, he or she did not care whether or not he or she violated the standard.[7]

According to these general ideas, we have the following order of levels of responsibility, with respect to the i agent's representation of what he or she has done:

- 1. i knows what he has done.

- 2. i ignores what he has done.

- 3. i's belief about what he has done is wrong.

We have a similar order for i agent's representation of what the other agents do:

- A. i knows what they have done.

- B. i ignores what they have done.

- C. i's belief about what they have done is wrong.

3.2.1 Obligation violation

We have seen that an obligation of the form $Obg(JBri_{Act}(\phi))$ is violated in the three situations formally represented by:

Violation 1: $(\neg Done_{i:a}) \wedge (Done_{Act-\{i:a\}})$.

Violation 2: $(\neg Done_{i:a}) \wedge (\neg Done_{Act-\{i:a\}})$.

Violation 3: $(Done_{i:a}) \wedge (\neg Done_{Act-\{i:a\}})$.

The intuitive meaning of the notation: $\phi \gg \psi$, is that the level of responsibility of i is stronger in a situation represented by ϕ than in a situation represented by ψ.

---

[7] Note that i may have performed another $i : b$ act. Here we focus on the responsibility of i for the fact that it did a. This responsibility should not be confused with i's responsibility for the fact that he did b.

Case of violation 1. Since we have $(\neg Done_{i:a})$, situations of types 1, 2 and 3 are respectively represented by $Bel_i(\neg Done_{i:a})$, $Ign_i(Done_{i:a})$ and $Bel_i(Done_{i:a})$. Then, we have:

(F1) $Bel_i(\neg Done_{i:a}) \gg Ign_i(Done_{i:a}) \gg Bel_i(Done_{i:a})$

We also have $(Done_{Act-\{i:a\}})$ and situations of types A, B and C are respectively represented by $Bel_i(Done_{Act-\{i:a\}})$, $Ign_i(Done_{Act-\{i:a\}})$ and $Bel_i(\neg Done_{Act-\{i:a\}})$. Then, we have:

(G1) $Bel_i(Done_{Act-\{i:a\}}) \gg Ign_i(Done_{Act-\{i:a\}}) \gg Bel_i(\neg Done_{Act-\{i:a\}})$

The different cases of representation of the situation by i are presented in Table Tab1.

Tab_1	$Bel_i(Done_{i:a})$	$Ign_i(Done_{i:a})$	$Bel_i(\neg Done_{i:a})$
$Bel_i(Done_{Act-\{i:a\}})$	l6	l3	max1
$Ign_i(Done_{Act-\{i:a\}})$	l7	l4	l1
$Bel_i(\neg(Done_{Act-\{i:a\}}))$	min1	l5	l2

From (F1) we can infer that each level of the set $\{max1, l1, l2\}$ is greater than each level of the set $\{l3, l4, l5\}$, and each level of the set $\{l3, l4, l5\}$ is greater than each level of the set $\{l6, l7, min1\}$.

In a similar way, from (G1) we can infer that each level of the set $\{max1, l3, l6\}$ is greater than each level of the set $\{l1, l4, l7\}$, and each level of the set $\{l1, l4, l7\}$ is greater than each level of the set $\{l2, l5, min1\}$.

Finally, we have a partial order that can be represented by:

$$max1 \gg l1 \gg l3 \gg l6 \gg l7 \gg min1$$

$$l3 \gg l4 \gg l5$$

$$max1 \gg l1 \gg l2 \gg l5 \gg min1$$

$$l1 \gg l4 \gg l7$$

Case of violation 2. Since we have: $(\neg Done_{i:a}) \wedge (\neg Done_{Act-\{i:a\}})$, the application of the same ordering criteria lead to:

(F1) $Bel_i(\neg Done_{i:a}) \gg Ign_i(Done_{i:a}) \gg Bel_i(Done_{i:a})$, and

(G2) $Bel_i(\neg Done_{Act-\{i:a\}}) \gg Ign_i(Done_{Act-\{i:a\}}) \gg Bel_i(Done_{Act-\{i:a\}})$

The different cases of i's representation of the situation are shown in table Tab2.

Tab_2	$Bel_i(Done_{i:a})$	$Ign_i(Done_{i:a})$	$Bel_i(\neg Done_{i:a})$
$Bel_i(Done_{Act-\{i:a\}})$	min2	m5	m2
$Ign_i(Done_{Act-\{i:a\}})$	m7	m4	m1
$Bel_i(\neg(Done_{Act-\{i:a\}}))$	m6	m3	max2

In a similar way, it can be shown that we have the partial order:

$$max2 \gg m3 \gg m6 \gg m7 \gg min2$$

$$m1 \gg m4 \gg m7$$

$$max2 \gg m1 \gg m2 \gg m5 \gg min2.$$

$$m3 \gg m4 \gg m5$$

Case of violation 3. Since we have $(Done_{i:a})$, agent i is not responsible of the obligation violation.

3.2.2 Prohibition violation

A prohibition is violated when we have $(Done_{i:a} \wedge Done_{Act-\{i:a\}})$. Since we have $(Done_{i:a})$, situations of types 1, 2 and 3 are represented by:

(F3) $Bel_i(Done_{i:a}) \gg Ign_i(Done_{i:a}) \gg Bel_i(\neg Done_{i:a})$.

Since we have $(Done_{Act-\{i:a\}})$ situations of type A, B and C are represented by:

(G3) $Bel_i(Done_{Act-\{i:a\}}) \gg Ign_i(Done_{Act-\{i:a\}}) \gg Bel_i(\neg Done_{Act-\{i:a\}})$.

The different cases of i's representation of the situation are shown in table Tab3.

Tab_3	$Bel_i(Done_{i:a})$	$Ign_i(Done_{i:a})$	$Bel_i(\neg Done_{i:a})$
$Bel_i(Done_{Act-\{i:a\}})$	max3	n3	n6
$Ign_i(Done_{Act-\{i:a\}})$	n1	n4	n7
$Bel_i(\neg(Done_{Act-\{i:a\}}))$	n2	n5	min3

Here, the partial order on responsibility levels is represented by:

$$max3 \gg n3 \gg n6 \gg n7 \gg min3$$

$$n3 \gg n4 \gg n5$$

$$max3 \gg n1 \gg n2 \gg n5 \gg min3.$$

$$n1 \gg n4 \gg n7$$

3.3   Responsibility in the case of a set of sets of acts

In the case of sets of sets of acts the formal analysis of responsibilities is more complex. That, is why we do not present such a detailed analysis. Nevertheless, the general principles of responsibility ordering are the same.

The obligations and prohibitions are represented by the formulas:

- obligation $Obg(SJBri_{Act}(\phi))$
- prohibition $Proh(SJBri_{Act}(\phi))$

Obligation violation This obligation is violated in the situations where we have: $\neg(Done_{Act_1} \wedge \ldots \wedge Done_{Act_n})$.

where the Act_is are sets of acts and: $Act^* = \{Act_1, \ldots, Act_n\}$.

It is assumed that agent i is acting in Act_1. Then, the violation situations can be equivalently represented by: $\neg(Done_{Act_1} \wedge Done_{Act^*-Act_1})$, which is logically equivalent to:

$\neg(Done_{Act_1}) \vee \neg(Done_{Act^*-Act_1})$

Therefore the set of all violation situations can be represented by the disjunction of the formulas:

Violation V1: $(\neg Done_{Act_1}) \wedge (Done_{Act^*-Act_1})$.

Violation V2: $(\neg Done_{Act_1}) \wedge \neg(Done_{Act^*-Act_1})$.

Violation V3: $(Done_{Act_1}) \wedge \neg(Done_{Act^*-Act_1})$.

Case of violation V1. In that case the sets of acts $Act^* - Act_1$ fulfill the obligation and $Obg(SJBri_{Act}(\phi))$ is fulfilled if the obligation $Obg(JBri_{Act_1}(\phi))$ is fulfilled.

Then i's responsibility levels can be analyzed in the same way as in the previous sub-section. The difference is that if i's belief is represented by $Bel_i(\neg Done_{Act_1})$ i may believe that his attitude cannot influence the global violation, and his responsibility levels are lower than in the cases where i's belief is represented by $Bel_i(Done_{Act_1})$.

Case of violation V2. The main difference from the V1 case is that Act_1 is not the only cause of the violation. Then, i's responsibility levels are lower than in V1.

Case of violation V3. In that case Act_1 has fulfilled the obligation $Obg(JBri_{Act_1}(\phi))$. Then, i is not responsible of $Obg(SJBri_{Act}(\phi))$ violation.

Tab_4	$Bel_i(Done_{Act_1})$	$Bel_i(\neg Done_{Act_1})$
$Bel_i(Done_{Act^*-Act_1})$	B4	B1
$Bel_i(\neg(Done_{Act^*-Act_1}))$	B3	B2

Prohibition violation This prohibition is violated in the situations where we have: $Done_{Act_1} \wedge \ldots \wedge Done_{Act_n}$.

If i's beliefs are represented by B1 or B4 (see table Tab_4), i knows that if $Done_{Act_1}$ has been performed the prohibition has been violated. In the case of B4 i's responsibility is stronger than in the case of B1 because i knows that the prohibition $Proh(JBri_{Act_1}(\phi))$ has been violated.

If i's beliefs are represented by B2 or B3, i believes that the prohibition was not violated, whether or not $Done_{Act_1}$ was executed. Then, his responsibility is lower than in the previous case because i believes that his attitude cannot influence the fact that the prohibition has been violated or not.

4   Related works

Braham and van Hees in their paper [3] analyse similar definitions of causality as the definitions presented in this paper. However, the motivation of this work is different, it is about game theory. The main difference is that definitions are not expressed in formal logic and no formal properties are derived from these definitions.

Pauly has defined in [21] a Coalition Logic to represent groups of agents who are acting together (see also [1]). This logic can be seen as an extension of Harel's Dynamic Logic [11] to multiple agents. Other extensions are presented in [4]. The common feature of these logics is that they accept a property of the form $[G]\top$, where G is a group of agents, which clearly shows that they do not represent causality — since a group of agents cannot bring it about that a tautology is true.

In [15] (see also [16]) Horty has defined an action operator to represent the fact that a group of agents "sees to it that" ϕ is the case. This operator is usually abbreviated as a "STIT" operator. The $[G \; cstit : \phi]$ operator indicates that the group of agents G sees to it that ϕ.

We have shown in [9] that this operator does not exactly represent causality. For instance, if in a situation where a door is open an agent can choose either to close the door or to do nothing. According to the definition of the $STIT$ operator, when the agent does nothing we can infer that he sees to it that the door is open. However, it is counterintuitive to say that doing nothing can cause (according to the meaning of causality) a given state of affairs.

In [7, 6] (see also [23]) Carmo has extended this operator to the $[G \; dstit : \phi]$ operator, which means that G deliberatively sees to it that ϕ. However, we have seen before that this operator does not really formalize causality.

In [20] de Sousa and Carmo presents an analysis of causality when the outcome is obtained by the actions of several agents acting successively. Causality when an agent acts alone is defined as usual by the two conditions that express that the action is sufficient and necessary (counterfactual condition). The systematic analysis of specific examples is not based on a formalization in logic. It is based on a temporal tree structure where, after each node in the tree, there are two branches that correspond to the case where the next action is performed or not. This article clearly shows, thanks to the examples, the problems that lawyers have in defining causality when several agents act successively. The main difference with what is presented here is that we consider several agents acting simultaneously, not successively, and that the problems are expressed in modal logic.

In [19] Lorini and Schwarzentruber have used the $[G \ cstit : \phi]$ operator[8] defined by Horty to formalize different kinds of counterfactual emotions. In [13] Hilpinen has defined a *"necessitating agency"* operator D (see AD9 in [13]).

According to its truth conditions, $D\phi$ is true in a world w'[9] iff there exists a world w and an action α such that 1) $w' \in g(\alpha, w)$ and ϕ is true in w', and 2) there exists a world w'' such that $< w, w'' >$ is *"maximally similar to the course of action exemplified by $< w, w' >$"* and ϕ is false in w''. In condition 1) $g(\alpha, w)$ denotes the set of worlds where we are after performing an action of type α (and possibly other actions). There is a strong similarity between a tuple of worlds $< w, w', w'' >$ that satisfies these conditions and a tuple that satisfies $R_{All, \{i:\alpha\}}(w, w', w'')$, for some $i : \alpha$, in the framework presented in this paper.

It is worth noting that in the definition of the $STIT$ operator we do not have a similar ternary relationship which allows to make explicit that a counterfactual world w'' is related to the performance of a given instance of a given action type denoted by w'.

In [25] Sergot has proposed formal definitions for joint actions operators in contexts where agents are acting collectively. These definitions are inspired by the definition of Pörn's bringing it about operator. However there is a significant difference. These operators are intended to characterize *how* the agents are acting, and not the final state of affairs which is obtained.

5 Conclusion

We have proposed a definition of the $JBri$ joint action operator to represent situations where the set of acts performed by a group of agents is sufficient to obtain a situation where a proposition ϕ holds, and all the acts in that set are necessary to obtain that situation.

---

[8]Notation has been modified here to simplify the comparison with previous work.
[9]Notation has been modified here to simplify the comparison with previous work.

The case study presented in the Introduction shows that the definition of causality formally represented by the joint actions operator $JBri$ has to be modified in case several sets of agents independently caused that ϕ holds, in the sense that each set of agents, if they acted alone, would have caused ϕ. To represent causality in these kinds of situations we have defined the restricted joint operators $RJBri$ and $SJBri$.

In addition to the formal definitions of these operators, we have formally shown some of their logical properties in Theorems 1 to 5.

Next, we analyzed the level of responsibility of an agent when acting in a group of agents that has violated an obligation or prohibition. We have shown that a wide range of situations must be considered if we accept that the levels of responsibility depend on the acts performed by the agent and other agents, and also on what the agent thinks of these acts.

Although the case study presented here is an academic example, we could easily find many similar examples in computer science. For example, in a context where it is forbidden to communicate a given password, two agents may have jointly informed another agent of the password by acting as follows: the first agent communicates the beginning of the password, and the second agent communicates the end of the password. If both agents have communicated the entire password simultaneously, the problem of assigning responsibilities is not insignificant.

Another example may be that it is forbidden to remove all copies of a given file. If there are only two copies in two different locations and two agents simultaneously launch a command to remove one copy each, they have removed all copies by their common actions and this was achieved indirectly (since each command caused only one of several software agents to execute).

The work presented here could be extended in several directions. One of them is to find a complete axiomatization of what has been defined in the semantics.

Acknowledgements

We want to thank the reviewers for their very helpful comments. We especially want to thank the reviewer who suggested many improvements and pointed out several errors. Thanks also to Hélène Bellières and Samson Bellières for their help in writing in English.

Annex

Proof of (¬DM). We can define a model M and a world w such that $M, w \models JBri_{Act[All]}(\phi \wedge \psi)$ and there exists $i : \alpha$ in Act such that $R_{All,\{i:\alpha\}}(w, w_1', w_1'')$ and (1) $M, w_1'' \models \phi \wedge \neg\psi$. Then, we have: $M, w_1'' \models \neg(\phi \wedge \psi)$. Since it is assumed that there is more than one act in Act we can also define $j : \beta$ in Act such that $R_{All,\{j:\beta\}}(w, w_2', w_2'')$ and (2) $M, w_2'' \models \psi \wedge \neg\phi$. Then, we have: $M, w_2'' \models \neg(\phi \wedge \psi)$.

In w_1'' we have $\phi \wedge \neg\psi$, which is not an inconsistent formula because it is assumed that we have $\not\models \phi \rightarrow \psi$. In the same way in w_2'' we have $\neg\phi \wedge \psi$, which is not an inconsistent formula because it is assumed that $\not\models \psi \rightarrow \phi$. From (1) we can infer $M, w \models \neg JBri_{Act[All]}(\phi)$ because the condition 2) in the Definition 4 is not satisfied. For the same reason, from (2) we can infer $M, w \models \neg JBri_{Act[All]}(\psi)$.

Finally we have: $M, w \models JBri_{Act[All]}(\phi \wedge \psi) \wedge \neg JBri_{Act[All]}(\phi) \wedge \neg JBri_{Act[All]}(\psi)$. Therefore we have: $\not\models JBri_{Act[All]}(\phi \wedge \psi) \rightarrow (JBri_{Act[All]}\phi \vee JBri_{Act[All]}\psi)$. End of proof.

The intuitive idea of the proof is that when $JBri_{Act[All]}(\phi \wedge \psi)$ holds, in all the counterfactual worlds w'' we have $\neg(\phi \wedge \psi)$, that is either ϕ or ψ is false. That is consistent with the fact that in some counterfactual worlds, like w_1'', ϕ is true, and w_1'' is not a counterfactual world for $JBri_{Act[All]}\phi$. In a similar way w_2'' is not a counterfactual world for $JBri_{Act[All]}\psi$.

References

[1] T. Agotnes, W. van der Hoek, and M. Wooldridge. On the logic of coalitional games. In *Proceedings of the Conference on Autonomous Agents and Multi Agent Systems*. Association for Computing Machinery, 2006.

[2] L. Aqvist. Old foundations for the logic of agency and action. *Studia Logica*, 72, 2002.

[3] M. Braham and M. Van Hees. An Anatomy of Moral Responsibility. *Mind*, (121), 2012.

[4] J. Broersen, A. Herzig, and N. Troquard. Normal Coalition Logic and its conformant extension. In D. Samet, editor, *Theoretical Aspects of Rationality and Knowledge*. Presses Universitaires de Louvain, 2007.

[5] J. Broersen. On the reconciliation of logics of agency and logics of event types. In K. Segerberg, editor, *On Logic of Actions*. Springer, 2014.

[6] J. Carmo. Collective agency, direct action and dynamic operators. *Logic Journal of the IGPL*, 18(1):66–98, 2010.

[7] J. Carmo and O. Pacheco. Deontic and action logics for organized collective agency, modeled through institutionalized agents and roles. *Fundamenta Informaticae*, 48:129–163, 2001.

[8] R. Demolombe. Relationships between obligations and actions in the context of institutional agents, human agents or software agents. *Journal of Artificial Intelligence and Law*, 19(2), 2011.

[9] R. Demolombe. Causality in the context of multiple agents. In T. Agotnes and J. Broersen and D. Elgesem, editor, *Deontic Logic in Computer Science, LNAI 7393*. Springer Verlag, 2012.

[10] R. Demolombe and A.J. Jones. Actions and normative positions. A modal-logical approach. In D. Jacquette, editor, *Companion to Philosophical Logic*. Blackwell, 2002.

[11] D. Harel. Dynamic logic. In D. Gabbay and F. Guenthner, editors, *Handbook of Philosophical Logic*, volume 2. Reidel, 1984.

[12] R. Hilpinen. *Deontic Logic : Introductory and Systematic Readings (edited)*. D. Reidel, 1971.

[13] R. Hilpinen. On Action and Agency. In E. Ejerhed and S. Lindstrom, editors, *Logic, Action and Cognition: Essays in Philosophical Logic*. Kluwer, 1997.

[14] R. Hilpinen. Las acciones como transformaciones de estado en g. h. von wright. *Doxa. Cuadernos de Filosofía del Derecho*, 2016.

[15] J. Horty. *Agency and deontic logic*. Oxford University Press, 2001.

[16] J.F. Horty and N. Belnap. The deliberative STIT: a study of action, omission, ability, and obligation. *Journal of Philosophical Logic*, 24:583–644, 1995.

[17] D. Lewis. *Counterfactuals*. Harvard University Press, 1973.

[18] L. Lindahl. *Position and Change – A Study in Law and Logic*. Synthese Library 112, D. Reidel, 1977.

[19] E. Lorini and F. Schwarzentruber. A logic for reasoning about counterfactual emotions. *Artificial Intelligence*, 175:814–847, 2011.

[20] P. de Sousa Mendes J. Carmo. A semantic model for causation in criminal law and the need of logico-legal criteria for the attribution of causation. *Law, Probability and Risk*, 12(3-4):207–228, 2013.

[21] M. Pauly. A modal logic for coalitional power in games. *Journal of Logic and Computation*, 12(1):149–166, 2002.

[22] I. Pörn. Action Theory and Social Science. Some Formal Models. *Synthese Library*, 120, 1977.

[23] F. Santos and J. Carmo. Indirect Action, Influence and Responsibility. In M. Brown and J. Carmo, editors, *Deontic Logic, Agency and Normative Systems, Workshops in Computing Series*. Springer, 1996.

[24] K. Segerberg. Outline of a logic of action. In F. Wolter, H. Wansing, W. de Rijke, and M. Zakharyaschev, editors, *Advances in Modal Logic, Volume 3*. World Scientific Publishing Co., 2002.

[25] M. Sergot. The logic of unwitting collective agency. Technical report 2088/6, Imperial College, London, 2008.

[26] G. H. von Wright. *Norm and Action*. Routledge and Kegan, 1963.

Received 1 September 2021

States and Internal States on Ehoops

Fei Xie

School of Mathematics and Statistics, Shandong Normal University, 250014, Jinan, P. R. China
850938132@qq.com

Hongxing Liu*

School of Mathematics and Statistics, Shandong Normal University, 250014, Jinan, P. R. China
lhxshanda@163.com

Abstract

An Ehoop is a generalization of hoops, where the top element is not guaranteed. In this paper, we study states and internal states on Ehoops with a bottom element. We present the notions of Bosbach states and Riečan states on Ehoops with a bottom element, and derive that these two kinds of states are consistent. It is shown that every Ehoop with a bottom element admits a Bosbach/Riečan state. Moreover, we investigated internal states on Ehoops with a bottom element. Also, prime state ideal theorem is given. Using prime state ideals, we establish a topological space.

Keyword: Ehoop, Bosbach state, Riečan state, State-morphism, Internal state, Prime state ideal.

1 Introduction

Hoops are basic residuated algebraic structures, which were introduced by Büchi and Owens in [7]. MV-algebras and BL-algebras are special cases of hoops. Recently, many researchers have studied hoops and derived many results. For example, many works on filters of hoops were in [21, 24, 25]. Also, Borzooei and Aaly Kologani ([3]) investigated right, left and product stabilizers on hoops. In addition, Aaly Kologani and Borzooei ([1]) introduced the notion of ideals in hoops, as a dual notion

We are very grateful to the referees for their valuable suggestions for improving this paper.

*Corresponding Author.

of filters. They studied implicative (maximal, prime) ideals and the relationships between these ideals in hoops.

Dvurečenskij and Zahiri ([16]) introduced EMV-algebras, which are generalizations of MV-algebras and generalized Boolean algebras. The top element of an EMV-algebra is not assumed. It was said that each EMV-algebra can be embedded into an EMV-algebra with a top element. Liu ([22]) presented the notion of EBL-algebras, as extensions of BL-algebras and EMV-algebras. It was proved that the set of all ideals of an EBL-algebra is in a one-to-one correspondence with the set of all congruences on an EBL-algebra. Also, under some condition, every EBL-algebra can be embedded into an EBL-algebra with a top element. In [27], we introduced Ehoops as a generalization of hoops. An Ehoop locally resembles hoops, which does not necessarily have a top element. In Ehoops, disjunction and multiplication exist but implication exists only in a local sense. It is known that a hoop is equivalent to an Ehoop with a top element and vice-versa. We investigated filters and ideals in Ehoops and defined congruences by ideals and filters respectively. When an Ehoop A satisfies the double negation property, the set of all ideals of A are in a one-to-one correspondence with the set of all congruences on A. Since the top element is not guaranteed for each Ehoop, there are some difficulties in the study of filters of an Ehoop.

A state plays a great role in the theory of quantum structures. Therefore, it is a natural problem to study states in many kinds of fuzzy structures. We note that states were studied in many different algebras, such as MV-algebras ([23]), BL-algebras ([26]), $R\ell$-monoids ([15]) and their non-commutative cases. Recently, in [17], Dvurečenskij and Zahiri investigated states on EMV-algebras.

As we all know, pseudo-hoops and semihoops are extensions of hoops. Some researchers also studied states in these algebraic structures. Ciungu ([8]) studied Bosbach and Riečan states on pseudo-hoops. It was proved that each Bosbach state on a good pseudo-hoop is a Riečan state and the two kinds of states are the same thing on a bounded Wajsberg pseudo-hoop. Bosbach states on subinterval algebras of a pseudo-hoop were also studied. Later, Ciungu in [9] studied state pseudo-hoops and state-morphism pseudo-hoops. In addition, Ciungu and Kühr ([11]) investigated the generalized states in bounded pseudo-BCK algebras and bounded pseudo-hoops. He, Zhao and Xin introduced the notions of Bosbach states, Riečan states and internal states on bounded semihoops in [20]. They showed that Bosbach states are Riečan states on bounded semihoops but the converse is not true in general. When the semihoops satisfies Glivenko property, the two kinds of states coincide. They also established a topological space applying prime state filters. In order to further study Ehoops, we shall introduce the notions of states and internal states on Ehoops and investigate their properties.

This paper is organized as follows. In Sect. 2, we gather some notions and results to be used. In Sect. 3, we investigate Bosbach states and Riečan states on Ehoops with a bottom element. State-morphisms on Ehoops with a bottom element are also studied. Applying the relationship between state-morphisms and maximal filters of an Ehoop with a bottom element, we prove that each Ehoop with a bottom element admits at least one Bosbach/Riečan state. We introduce the notion of internal states on an Ehoop with a bottom element in Sect. 4. In addition, some properties of state ideals and prime state ideals on state Ehoops are given. Applying prime state ideals on state Ehoops, we establish the topological structures.

2 Preliminaries

In this section, we shall recall some notions and results on hoops and Ehoops.

Definition 2.1. [2, 7] A hoop is an algebra $(A, \odot, \rightarrow, 1)$ of type $(2, 2, 0)$ with the following axioms for any $x, y, z \in A$:

(i) $(A, \odot, 1)$ is a commutative monoid;
(ii) $x \rightarrow x = 1$;
(iii) $(x \odot y) \rightarrow z = x \rightarrow (y \rightarrow z)$;
(iv) $x \odot (x \rightarrow y) = y \odot (y \rightarrow x)$.

A hoop is said to be bounded if there is a bottom element 0. In this case, we write x^- instead of $x \rightarrow 0$ for any $x \in A$. A bounded hoop A satisfies the double negation property if $x^{--} = x$ for all $x \in A$.

For any semigroup $(A, \odot)$, we say that $x \in A$ is idempotent if $x \odot x = x$. Use $Id(A)$ to denote the set $\{x \in A | x \odot x = x\}$. Due to [14, Proposition 3.1], we note that the element b of a hoop A is idempotent if and only if $x \odot b = x \wedge b$ for any $x \in A$.

The following are some properties of hoops which will be used in this paper.

Proposition 2.2. [4, 5, 10, 18] Let A be a hoop. For all $x, y, z \in A$, we have:

(i) $x \leq y \rightarrow z \Leftrightarrow x \odot y \leq z$;
(ii) $1 \rightarrow y = y$;
(iii) $x \odot y \leq x \wedge y = x \odot (x \rightarrow y)$;
(iv) if $x \leq y$, then $x \odot z \leq y \odot z$, $y \rightarrow z \leq x \rightarrow z$ and $z \rightarrow x \leq z \rightarrow y$;
(v) $x \rightarrow y \leq (y \rightarrow z) \rightarrow (x \rightarrow z)$, $x \rightarrow y \leq (z \rightarrow x) \rightarrow (z \rightarrow y)$;
(vi) $y \rightarrow (x \rightarrow z) = x \rightarrow (y \rightarrow z)$;
(vii) $z \rightarrow (x \wedge y) = (z \rightarrow x) \wedge (z \rightarrow y)$;
(viii) if A is bounded, then $x \odot x^- = 0$, $x \leq x^{--}$ and $x^{---} = x^-$;

(ix) *if A is bounded, then $x \odot y = 0 \Leftrightarrow x \leq y^{-}$;*

(x) *if A is bounded, then $(y \to x)^{--} = y^{--} \to x^{--}$.*

Definition 2.3. *[27] An Ehoop is an algebra $(A, \wedge, \odot)$ of type $(2, 2)$ such that:*

(i) *$(A, \wedge)$ is a $\wedge$-semilattice;*

(ii) *$(A, \odot)$ is a commutative semigroup;*

(iii) *for all $a \in Id(A)$, $x \to_a y = max\{z \in A_a | x \odot z \leq y\}$ exists for any $x, y \in A_a$, and $(A_a, \odot, \to_a, a)$ is a hoop, where $A_a = \{x \in A | x \leq a\}$;*

(iv) *there are enough idempotent elements in A, that is, for all $x, y \in A$, there exists $a \in Id(A)$ with $x, y \leq a$.*

Let A be an Ehoop. Define $x^n = x^{n-1} \odot x$ for all $x \in A$, where the integer n satisfies $n \geq 2$. If $a \in Id(A)$, we define $x \leq_a y \Leftrightarrow x \to_a y = a$ for any $x, y \in A_a$. Then $\leq_a$ is a partial order on A_a. For any $x, y \in A$, any $a, b \in Id(A)$ with $x, y \leq a$ and $x, y \leq b$, we get that $x \leq_a y \Leftrightarrow x \leq y \Leftrightarrow x \wedge y = x \Leftrightarrow x \leq_b y$ and $x \wedge y = x \odot (x \to_a y) = x \odot (x \to_b y)$.

Suppose that there is a bottom element 0 in A. If $a \in Id(A)$, we define $x^{-a} = x \to_a 0$ and $x \ominus_a y = x^{-a} \to_a y$ for any $x, y \in A_a$. For any $x, y \in A$, if $(x \odot y)^{-a-a} = x^{-a-a} \odot y^{-a-a}$ for all $a \in Id(A)$ with $x, y \leq a$, we say that the Ehoop A is normal. If A_a satisfies the double negation property for all $a \in Id(A)$, A is an Ehoop with the double negation property. It is clear that an Ehoop with the double negation property is normal.

Proposition 2.4. *[27] Let A be an Ehoop. If $a, b \in Id(A)$ and $a \leq b$, the following properties hold for any $x, y, z \in A_a$:*

(i) *$x \to_a y = (x \to_b y) \wedge a$;*

(ii) *$(x \to_a y) \to_a z \leq (x \to_b y) \to_b z$.*

Proposition 2.5. *[27] Let A be an Ehoop with a bottom element 0. The following properties hold for all $a \in Id(A)$, for any $x, y, z, z_i \in A_a$:*

(i) *$x, y \leq x \ominus_a y$;*

(ii) *if $x \leq y$, then $x \ominus_a z \leq y \ominus_a z$ and $z \ominus_a x \leq z \ominus_a y$;*

(iii) *if A is normal, $\ominus_a$ is associative;*

(iv) *if A satisfies the double negation property, $\ominus_a$ is commutative;*

(v) *if A satisfies the double negation property, $x \ominus_a (\wedge_{i \in I} z_i) = \wedge_{i \in I}(x \ominus_a z_i)$, when $\wedge_{i \in I} z_i$ exists;*

(vi) *if A satisfies the double negation property, $x \wedge (z_1 \ominus_a z_2 \ominus_a \cdots \ominus_a z_n) \leq (x \wedge z_1) \ominus_a (x \wedge z_2) \ominus_a \cdots \ominus_a (x \wedge z_n)$.*

Let A be an Ehoop with a bottom element 0. When A is normal, we define $2_a y = y \ominus_a y$, $3_a y = y \ominus_a (y \ominus_a y)$, $\cdots$, $n_a y = y \ominus_a (n-1)_a y$, where $y \in A$ and $a \in Id(A)$ such that $y \leq a$. Also, define $\ominus_{a}{}^{n}_{i=1} y_i = y_1 \ominus_a y_2 \ominus_a \cdots \ominus_a y_n$ for any $y_1, y_2, \cdots, y_n \in A_a$, where $n \in \mathbb{N}\backslash\{0\}$ and $a \in Id(A)$.

A subalgebra B of an Ehoop A is a subset of A with conditions: (i) B is closed under $\wedge$ and $\odot$, (ii) for any $b \in Id(A) \cap B$, $B_b = \{x \in B | x \leq b\}$ is a subalgebra of A_b, and (iii) for any $x, y \in B$, there exists $b \in Id(A) \cap B$ with $x, y \leq b$.

A function $f : A_1 \to A_2$ between two Ehoops A_1 and A_2 is called an Ehoop homomorphism if f preserves $\wedge$ and $\odot$, and for all $a \in Id(A)$ and for any $x, y \in A_a$, we have $f(x \to_a y) = f(x) \to_{f(a)} f(y)$.

Let A be an Ehoop. A congruence θ on A is an equivalence relation satisfying the conditions: (i) θ is compatible with $\wedge$ and $\odot$, and (ii) $\theta \cap (A_a \times A_a)$ is a congruence on A_a for any $a \in Id(A)$. If θ is a congruence on A, define $A/\theta = \{x/\theta \mid x \in A\}$, where $x/\theta = \{y \in A \mid (x,y) \in \theta\}$. It is shown that the quotient algebra $(A/\theta, \wedge, \odot)$ is an Ehoop, where

$$x/\theta \wedge y/\theta = (x \wedge y)/\theta \quad \text{and} \quad x/\theta \odot y/\theta = (x \odot y)/\theta,$$

for all $x/\theta, y/\theta \in A/\theta$. In addition, for any $x/\theta, y/\theta \in (A/\theta)_{a/\theta}$ with $a \in Id(A)$ and $x, y \leq a$, we have $x/\theta \to_{a/\theta} y/\theta = (x \to_a y)/\theta$.

Let A be an Ehoop with a bottom element 0. An ideal I of A is a subset of A which satisfies: (i) if $x, y \in I$, then $x \ominus_a y \in I$ for all $a \in Id(A)$ with $x, y \leq a$, and (ii) if $x \leq y$ and $y \in I$, then $x \in I$.

A filter F of A is a non-empty subset of A satisfying the conditions: (i) for any $x \in A$, there exists $a \in Id(A) \cap F$ with $x \leq a$, (ii) for any $x, y \in A$, $x \leq y$ and $x \in F$ imply $y \in F$, and (iii) F is closed under $\odot$. Equivalently, a non-empty subset F of A is a filter iff F satisfies that (i) for any $x \in A$, there exists $a \in Id(A) \cap F$ with $x \leq a$ and (ii) for any $x, y \in A$, for all $b \in Id(A)$ such that $x, y \leq b$, if $x, x \to_b y \in F$, then $y \in F$. Denote by $(X]$ the filter of A generated by the subset X of A. A filter F of A is called maximal if $(F \cup \{x\}] = A$ for all $x \in A \backslash F$. A filter F of A is said to be proper if $F \neq A$. Let F be a filter of A. Define the relation θ_F on A by $(x, y) \in \theta_F$ if and only if there is $a \in Id(A)$ with $x, y \leq a$ and $x \to_a y, y \to_a x \in F$. Then θ_F is a congruence on A. The corresponding quotient Ehoop A/θ_F is denoted by A/F.

Proposition 2.6. [27] *Let F be a filter of an Ehoop A. Then F is maximal iff for all $x \in A$, if $x \notin F$, then for any $y \in A$, there exist $n \in \mathbb{N}\backslash\{0\}$ and $b \in Id(A)$ with $x, y \leq b$ and $x^n \to_b y \in F$.*

Remark 2.7. *Let F be a filter of an Ehoop A with a bottom element 0. Then F is maximal iff for any $x \in A$, if $x \notin F$, there exist $n \in \mathbb{N}\backslash\{0\}$ and $b \in Id(A)$ with $x \leq b$ such that $x^n \to_b 0 \in F$.*

3   States on Ehoops

In this section, we give the notions of Bosbach states and Riečan states on Ehoops with a bottom element. It is shown that Bosbach states are consistent with Riečan states on Ehoops with a bottom element. In addition, we introduce the notion of state-morphisms on Ehoops with a bottom element. The relationships between state-morphisms and maximal filters on Ehoops with a bottom element are investigated. Finally, we show that each Ehoop with a bottom element admits a Bosbach state and a Riečan state.

Definition 3.1. *Let A be an Ehoop with a bottom element 0. A Bosbach state on A is a mapping $s : A \longrightarrow [0,1]$ satisfying the following conditions:*

(i)   *$s(0) = 0$;*
(ii) *there exists $x_0 \in A$ such that $s(x_0) = 1$;*
(iii) *$s(x) + s(x \rightarrow_a y) = s(y) + s(y \rightarrow_a x)$ for any $x, y \in A$, $a \in Id(A)$ with $x, y \leq a$.*

Remark 3.2. *If A has a top element 1, then it is a hoop. A Bosbach state on a bounded hoop A is a mapping $s : A \longrightarrow [0,1]$ such that $s(0) = 0$, $s(1) = 1$ and $s(x) + s(x \rightarrow_1 y) = s(y) + s(y \rightarrow_1 x)$ for any $x, y \in A$.*

Example 3.3. *Let A be the set $\{0, \frac{1}{n-1}, \frac{2}{n-1}, \cdots, \frac{n-2}{n-1}, 1, 2, 3, \cdots\}$ with the natural order, where the integer n satisfies $n \geq 2$. Define $\odot$ and $\wedge$ as follows:*

$$u \wedge v = min\{u, v\},$$

$$u \odot v = \begin{cases} (u + v - 1) \vee 0, & \text{if } u, v \in [0,1], \\ u, & \text{if } u \in [0,1], \ v \geq 2, \\ v, & \text{if } u \geq v \geq 2. \end{cases}$$

When $a = 1 \in Id(A)$, define $u \rightarrow_1 v = (v - u + 1) \wedge 1$ for any $u, v \in [0,1]$. If $a \in Id(A)$ with $a \geq 2$, for any $u, v \in A_a$, set

$$u \rightarrow_a v = \begin{cases} a, & \text{if } u \leq v, \\ v - u + 1, & \text{if } 0 \leq v < u \leq 1, \\ v, & \text{if } 0 \leq v \leq 1 < u, \\ v, & \text{if } 1 < v < u. \end{cases}$$

By [27, Example 3.6], $(A, \wedge, \odot)$ is an Ehoop. Define a mapping $s : A \longrightarrow [0,1]$ by

$$s(u) = \begin{cases} u, & \text{if } u \in [0,1], \\ 1, & \text{if } u > 1. \end{cases}$$

Then s is a Bosbach state on A.

Example 3.4. *Let $A_1 = \{0, a, b, c, d, 1\}$ be the hoop from [24, Example 3.2(ii)], where the operations are defined as follows:*

$\rightarrow$	0	a	b	c	d	1
0	1	1	1	1	1	1
a	c	1	b	c	b	1
b	d	a	1	b	a	1
c	a	a	1	1	a	1
d	b	1	1	b	1	1
1	0	a	b	c	d	1

$\odot$	0	a	b	c	d	1
0	0	0	0	0	0	0
a	0	a	d	0	d	a
b	0	d	c	c	0	b
c	0	0	c	c	0	c
d	0	d	0	0	0	d
1	0	a	b	c	d	1

Define a mapping $s_1 : A_1 \longrightarrow [0, 1]$ by

$$
s_1(x) = \begin{cases}
0, & x = 0, \\
\frac{1}{3}, & x = a, \\
\frac{5}{6}, & x = b, \\
\frac{2}{3}, & x = c, \\
\frac{1}{6}, & x = d, \\
1, & x = 1.
\end{cases}
$$

Let A_2 be the Ehoop from Example 3.3. Then $A_1 \times A_2$ is also an Ehoop. Define $s' : A_1 \times A_2 \longrightarrow [0, 1]$ by $s'(x, y) = \lambda s_1(x) + (1 - \lambda)s(y)$, where $(x, y) \in A_1 \times A_2$, λ is a real number such that $0 < \lambda < 1$ and s is the mapping in Example 3.3. We can check that s' is a Bosbach state on $A_1 \times A_2$. In particular, set $\lambda = \frac{1}{3}$. Then $s'(x, y) = \frac{1}{3}s_1(x) + \frac{2}{3}s(y)$ is a Bosbach state on $A_1 \times A_2$.

Next, we give some properties of Bosbach states on an Ehoop A with a bottom element.

Proposition 3.5. *Let s be a Bosbach state on an Ehoop A with a bottom element 0. For all $x, y, z \in A$, we have*

(i) *$s(x) \leq s(a)$ and $s(x^{-a}) = s(a) - s(x)$, for any $a \in Id(A)$ with $x \leq a$;*
(ii) *$s(x^{-a-a}) = s(x)$, for any $a \in Id(A)$ with $x \leq a$;*
(iii) *if $x \leq y$, then $s(x) \leq s(y)$;*
(iv) *$s(x \rightarrow_a y) = s(y \rightarrow_a x) \iff s(x) = s(y)$, for any $a \in Id(A)$ with $x, y \leq a$;*
(v) *$s(x^{-a-a} \rightarrow_a x) = s(a)$, for any $a \in Id(A)$ with $x \leq a$;*

(vi) $s(x \to_a y^{-a-a}) = s(x^{-a-a} \to_a y) = s(x \to_a y)$, for any $a \in Id(A)$ with $x, y \le a$;

(vii) $s(x^{-a} \to_a y^{-a}) = s(y \to_a x)$, for any $a \in Id(A)$ with $x, y \le a$;

(viii) $s(x \to_a (y^{-a} \to_a z^{-a})) = s(x \to_a (z \to_a y))$, for any $a \in Id(A)$ with $x, y, z \le a$;

(ix) $s(x \odot y) = s(a) - s(x \to_a y^{-a})$, for any $a \in Id(A)$ with $x, y \le a$.

Proof. (i) Since s is a Bosbach state on A, we have $s(x^{-a}) + s(x) = s(0) + s(0 \to_a x) = 0 + s(a)$. Thus, $s(x^{-a}) = s(a) - s(x)$. It follows from $s(x^{-a}) \ge 0$ that $s(x) \le s(a)$.

(ii) By (i), $s(x^{-a-a}) = s(a) - s(x^{-a}) = s(a) - (s(a) - s(x)) = s(x)$.

(iii) Let $x \le y$. For any $a \in Id(A)$ with $x, y \le a$, we get that $s(x) + s(a) = s(x) + s(x \to_a y) = s(y) + s(y \to_a x)$. Then $s(x) - s(y) = s(y \to_a x) - s(a)$. By (i), the inequality $y \to_a x \le a$ implies $s(y \to_a x) \le s(a)$ and so $s(x) - s(y) \le 0$.

(iv) It is straightforward.

(v) Applying (ii), we obtain

$$s(x^{-a-a} \to_a x) + s(x) = s(x^{-a-a} \to_a x) + s(x^{-a-a})$$
$$= s(x) + s(x \to_a x^{-a-a})$$
$$= s(x) + s(a),$$

which yields $s(x^{-a-a} \to_a x) = s(a)$.

(vi) As $y \le y^{-a-a}$, we get that $x \to_a y \le x \to_a y^{-a-a}$. Hence, $s((x \to_a y) \to_a (x \to_a y^{-a-a})) = s(a)$. The inequality $y^{-a-a} \to_a y \le (x \to_a y^{-a-a}) \to_a (x \to_a y)$ together with (iii) and (v) implies $s(a) = s(y^{-a-a} \to_a y) \le s((x \to_a y^{-a-a}) \to_a (x \to_a y)) \le s(a)$. Thus, $s((x \to_a y^{-a-a}) \to_a (x \to_a y)) = s(a)$. Then

$$s(x \to_a y^{-a-a}) + s(a) = s(x \to_a y^{-a-a}) + s((x \to_a y^{-a-a}) \to_a (x \to_a y))$$
$$= s(x \to_a y) + s((x \to_a y) \to_a (x \to_a y^{-a-a}))$$
$$= s(x \to_a y) + s(a).$$

It follows that $s(x \to_a y^{-a-a}) = s(x \to_a y)$. Similarly, we can show $s(x^{-a-a} \to_a y) = s(x \to_a y)$.

(vii) By Proposition 2.2(vi) and (vi), we have $s(x^{-a} \to_a y^{-a}) = s(y \to_a x^{-a-a}) = s(y \to_a x)$.

(viii) From (vii) we get that $s(y^{-a} \to_a z^{-a}) = s(z \to_a y)$. This together with (iv) and Proposition 2.2(v) implies that

$$s((y^{-a} \to_a z^{-a}) \to_a (z \to_a y)) = s((z \to_a y) \to_a (y^{-a} \to_a z^{-a})) = s(a).$$

Since $(y^{-a} \to_a z^{-a}) \to_a (z \to_a y) \le (x \to_a (y^{-a} \to_a z^{-a})) \to_a (x \to_a (z \to_a y)) \le a$, we have

$$
\begin{aligned}
s(a) &= s((y^{-a} \to_a z^{-a}) \to_a (z \to_a y)) \\
&\le s((x \to_a (y^{-a} \to_a z^{-a})) \to_a (x \to_a (z \to_a y))) \\
&\le s(a),
\end{aligned}
$$

which entails $s((x \to_a (y^{-a} \to_a z^{-a})) \to_a (x \to_a (z \to_a y))) = s(a)$. In a similar way, we can show $s((x \to_a (z \to_a y)) \to_a (x \to_a (y^{-a} \to_a z^{-a}))) = s(a)$. By (iv), $s(x \to_a (y^{-a} \to_a z^{-a})) = s(x \to_a (z \to_a y))$.

(ix) Obviously, we have $s(x \odot y) + s((x \odot y)^{-a}) = s(0) + s(0 \to_a (x \odot y)) = 0 + s(a)$. This implies $s(x \odot y) = s(a) - s((x \odot y)^{-a}) = s(a) - s(x \to_a y^{-a})$. $\square$

Theorem 3.6. *Let A be an Ehoop with a bottom element 0. If $s : A \longrightarrow [0,1]$ is a mapping such that $s(0) = 0$ and there exists $x_0 \in A$ with $s(x_0) = 1$, then the following conditions are equivalent:*

(i) *s is a Bosbach state;*
(ii) *for any $x, y \in A$, if $x \le y$, then $s(y \to_a x) = s(a) + s(x) - s(y)$, where $a \in Id(A)$ with $x, y \le a$;*
(iii) *for any $x, y \in A$, $s(y \to_a x) = s(a) + s(x \wedge y) - s(y)$, where $a \in Id(A)$ with $x, y \le a$.*

Proof. (i) $\implies$ (ii) Let s be a Bosbach state. Suppose $x \le y$ and $a \in Id(A)$ with $x, y \le a$. Then the equality $s(x) + s(a) = s(x) + s(x \to_a y) = s(y) + s(y \to_a x)$ implies $s(y \to_a x) = s(x) + s(a) - s(y)$.

(ii) $\implies$ (iii) Suppose that condition (ii) holds. Let $x, y \in A$ and $a \in Id(A)$ such that $x, y \le a$. By the assumption, we have

$$
\begin{aligned}
s(y \to_a x) &= s((y \to_a x) \wedge a) \\
&= s((y \to_a x) \wedge (y \to_a y)) \\
&= s(y \to_a (x \wedge y)) \quad \text{(by Proposition 2.2(vii))} \\
&= s(a) + s(x \wedge y) - s(y).
\end{aligned}
$$

(iii) $\implies$ (i) Let $x, y \in A$ and $a \in Id(A)$ such that $x, y \le a$. The condition (iii) implies that $s(x) + s(x \to_a y) = s(x) + s(a) + s(x \wedge y) - s(x) = s(a) + s(x \wedge y)$. Similarly, $s(y) + s(y \to_a x) = s(a) + s(x \wedge y)$. Hence, we have $s(x) + s(x \to_a y) = s(y) + s(y \to_a x)$, which completes the proof. $\square$

Proposition 3.7. *Let s be a Bosbach state on an Ehoop A with a bottom element 0. Then the following statements hold:*

(i) $Ker(s) = \{x \in A | s(x) = 1\}$ *is a filter of A;*

(ii) $(x, y) \in \theta_{Ker(s)} \Longleftrightarrow s(x) = s(y) = s(x \wedge y)$;

(iii) $(A/Ker(s), \odot, \rightarrow_{Ker(s)}, Ker(s))$ *is a hoop, where $Ker(s)$ is the top element of $A/Ker(s)$;*

(iv) *there exists a unique Bosbach state $\tilde{s}$ on the hoop $A/Ker(s)$ with $\tilde{s}(x/Ker(s)) = s(x)$ for any $x \in A$.*

Proof. (i) Since s is a Bosbach state, there exists $x_0 \in A$ such that $s(x_0) = 1$. For any $x \in A$, choose $b \in Id(A)$ such that $x, x_0 \leq b$. Then $s(b) = 1$. Therefore, for any $x \in A$, there exists $b \in Id(A) \cap Ker(s)$ such that $x \leq b$. Let $x, y \in A$ and $a \in Id(A)$ such that $x, y \leq a$. Suppose that $x \in Ker(s)$ and $x \rightarrow_a y \in Ker(s)$. As $x \odot y \leq x$, i.e. $x \leq y \rightarrow_a x$, we have $s(x) \leq s(y \rightarrow_a x)$ and so $s(y \rightarrow_a x) = 1$. It follows from $s(x) + s(x \rightarrow_a y) = s(y) + s(y \rightarrow_a x)$ that $s(y) = 1$, i.e. $y \in Ker(s)$. Therefore, $Ker(s)$ is a filter of A.

(ii) Suppose $(x, y) \in \theta_{Ker(s)}$. There exists $a \in Id(A)$ such that $x, y \leq a$ and $x \rightarrow_a y$, $y \rightarrow_a x \in Ker(s)$. By Proposition 3.5(iv), $s(x) = s(y)$. The inequality $x \rightarrow_a y \leq a$ implies $s(a) = 1$. It follows from Theorem 3.6 that $s(x \wedge y) = s(y \rightarrow_a x) - s(a) + s(y) = 1 - 1 + s(y) = s(y) = s(x)$.

Conversely, suppose $x, y \in A$ and $s(x) = s(y) = s(x \wedge y)$. For any $a \in Id(A)$ with $x, y \leq a$, we have $s(y \rightarrow_a x) = s(a) + s(x \wedge y) - s(y) = s(a)$ by Theorem 3.6. In a similar way, $s(x \rightarrow_a y) = s(a)$ and so $s(x \rightarrow_a y) = s(y \rightarrow_a x) = s(a)$. Since there is $x_0 \in A$ such that $s(x_0) = 1$, we have $b \in Id(A)$ such that $x, y, x_0 \leq b$. Then $s(b) = 1$ and so $s(x \rightarrow_b y) = s(y \rightarrow_b x) = s(b) = 1$, i.e. $x \rightarrow_b y$, $y \rightarrow_b x \in Ker(s)$. Thus, $(x, y) \in \theta_{Ker(s)}$.

(iii) Since $Ker(s)$ is a filter of A, there is $a \in Id(A) \cap Ker(s)$ such that $x \leq a$ for any $x \in A$. Then $x/Ker(s) \leq a/Ker(s)$. From (ii) it follows that $a/Ker(s) = \{z \in A | s(z) = s(a) = s(z \wedge a)\} = Ker(s)$. Thus, $Ker(s)$ is the top element in the Ehoop $A/Ker(s)$ and so $(A/Ker(s), \odot, \rightarrow_{Ker(s)}, Ker(s))$ is a hoop.

(iv) Clearly, $\tilde{s}(0/Ker(s)) = s(0) = 0$. Since there exists $x_0 \in A$ such that $s(x_0) = 1$, we have $\tilde{s}(x_0/Ker(s)) = s(x_0) = 1$. By Remark 3.2, it suffices to show that for any $x, y \in A$, $\tilde{s}(x/Ker(s)) + \tilde{s}(x/Ker(s) \rightarrow_{Ker(s)} y/Ker(s)) = \tilde{s}(y/Ker(s)) + \tilde{s}(y/Ker(s) \rightarrow_{Ker(s)} x/Ker(s))$. Let $x, y \in A$. There exists $b \in Id(A) \cap Ker(s)$ such that $x, y \leq b$. Hence, we have $b/Ker(s) = Ker(s)$ and so

$$\tilde{s}(x/Ker(s)) + \tilde{s}(x/Ker(s) \rightarrow_{Ker(s)} y/Ker(s)) = s(x) + \tilde{s}((x \rightarrow_b y)/Ker(s))$$
$$= s(x) + s(x \rightarrow_b y).$$

In a similar way, $\tilde{s}(y/Ker(s)) + \tilde{s}(y/Ker(s) \rightarrow_{Ker(s)} x/Ker(s)) = s(y) + s(y \rightarrow_b x)$. As $s(x) + s(x \rightarrow_b y) = s(y) + s(y \rightarrow_b x)$, $\tilde{s}$ is a Bosbach state. $\square$

Let A be an Ehoop with a bottom element 0 and $x, y \in A$. If there exists $a \in Id(A)$ such that $x, y \leq a$ and $y^{-a-a} \leq x^{-a}$, the two elements x, y are called orthogonal and we write $x \perp y$. Obviously, for any $x, y \in A$, we have $x \perp y$ iff $y \perp x$.

Proposition 3.8. *Let A be an Ehoop with a bottom element 0. The following conditions are equivalent for any $x, y \in A$:*

(i) $x \perp y$;
(ii) *for any $a \in Id(A)$ with $x, y \leq a$, we have $y^{-a-a} \leq x^{-a}$;*
(iii) *for any $a \in Id(A)$ with $x, y \leq a$, we have $x^{-a-a} \odot y^{-a-a} = 0$;*
(iv) $x \odot y = 0$.

Proof. (i) $\implies$ (ii) Suppose $x \perp y$. There exists $a \in Id(A)$ with $x, y \leq a$ and $y^{-a-a} \leq x^{-a}$. Hence, $x^{-a-a} \leq y^{-a-a-a} = y^{-a}$. Let b be an arbitrary idempotent element of A such that $x, y \leq b$. We shall show that $y^{-b-b} \leq x^{-b}$. Choose $c \in Id(A)$ with $a, b \leq c$. Due to Proposition 2.4(i), we have $x \leq x^{-a-a} \leq y^{-a} \leq y^{-c}$, which yields $x \leq y^{-c} \wedge b = y^{-b}$. It follows that $y^{-b} \to_c 0 \leq x \to_c 0$ in the hoop A_c. Therefore, $(y^{-b} \to_c 0) \wedge b \leq (x \to_c 0) \wedge b$, i.e. $y^{-b-b} \leq x^{-b}$.

(ii) $\implies$ (i) It is obvious.

(ii) $\iff$ (iii) For any $a \in Id(A)$ with $x, y \leq a$, we have that $y^{-a-a} \leq x^{-a} = x^{-a-a-a}$ iff $y^{-a-a} \odot x^{-a-a} = 0$.

(iii) $\implies$ (iv) Let $a \in Id(A)$ such that $x, y \leq a$. By (iii), $x \odot y \leq x^{-a-a} \odot y^{-a-a} = 0$. Thus, $x \odot y = 0$.

(iv) $\implies$ (i) Suppose $x \odot y = 0$. Let $a \in Id(A)$ such that $x, y \leq a$. We have $x \leq y^{-a}$ and so $y^{-a-a} \leq x^{-a}$. $\square$

Let A be an Ehoop with a bottom element 0. If $x \perp y$, we define $+_a$ by $x +_a y = x^{-a} \to_a y^{-a-a}$ on A, where a is an arbitrary idempotent element with $x, y \leq a$. Clearly, $x +_a y = (x^{-a} \odot y^{-a})^{-a} = y +_a x$.

Definition 3.9. *Let A be an Ehoop with a bottom element 0. A Riečan state is a mapping $s : A \longrightarrow [0, 1]$ satisfying the following conditions:*

(i) *there exists $x_0 \in A$ such that $s(x_0) = 1$;*
(ii) *if $x \perp y$, for any $a \in Id(A)$ with $x, y \leq a$ we have $s(x +_a y) = s(x) + s(y)$.*

Example 3.10. *We can check that the mapping s in Example 3.3 and the mapping s' in Example 3.4 are also Riečan states.*

Proposition 3.11. *Let A be an Ehoop with a bottom element 0 and s a Riečan state. Then*

(i) $s(0) = 0$;

(ii) *for any $x \in A$, $a \in Id(A)$ with $x \leq a$, we have $s(x) \leq s(a)$ and $s(x^{-a}) = s(a) - s(x)$;*

(iii) *for any $x \in A$, $a \in Id(A)$ with $x \leq a$, we have $s(x^{-a-a}) = s(x)$;*

(iv) *for any $x, y \in A$, $x \leq y$ implies $s(x) \leq s(y)$.*

Proof. (i) For any $a \in Id(A)$, we have $s(0 +_a 0) = s(0) + s(0)$. This together with $s(0 +_a 0) = s(0^{-a} \to_a 0^{-a-a}) = s(a \to_a 0) = s(0)$ implies $s(0) = s(0) + s(0)$. Thus, $s(0) = 0$.

(ii) Let $a \in Id(A)$ with $x \leq a$. Then $x^{-a} \perp x$ and $x^{-a} +_a x = x^{-a-a} \to_a x^{-a-a} = a$. It follows that $s(a) = s(x^{-a} +_a x) = s(x^{-a}) + s(x)$, which entails $s(x^{-a}) = s(a) - s(x) \geq 0$ and so $s(x) \leq s(a)$.

(iii) By (ii), $s(x^{-a-a}) = s(a) - s(x^{-a}) = s(a) - (s(a) - s(x)) = s(x)$.

(iv) Suppose $x \leq y$. For any $a \in Id(A)$ such that $x \leq y \leq a$, we obtain $y^{-a-a-a} = y^{-a} \leq x^{-a}$. That is, $x \perp y^{-a}$. Thus, $s(x +_a y^{-a}) = s(x) + s(y^{-a}) = s(x) + s(a) - s(y)$ and so $s(x) - s(y) = s(x +_a y^{-a}) - s(a)$. Since $x +_a y^{-a} \leq a$, we have $s(x +_a y^{-a}) \leq s(a)$ by (ii), which means $s(x) - s(y) \leq 0$. $\qquad\square$

The following theorem illustrates the relationship between Bosbach states and Riečan states on Ehoops with a bottom element.

Theorem 3.12. *Let A be an Ehoop with a bottom element 0. Then Bosbach states and Riečan states on A coincide.*

Proof. Suppose that s is a Bosbach state on A. There exists $x_0 \in A$ such that $s(x_0) = 1$. Let $x \perp y$. For any $a \in Id(A)$ with $x, y \leq a$, we have $y^{-a-a} \leq x^{-a}$, i.e. $y^{-a-a} \to_a x^{-a} = a$. By Proposition 3.5(i) and (ii), it follows from $s(x^{-a}) + s(x^{-a} \to_a y^{-a-a}) = s(y^{-a-a}) + s(y^{-a-a} \to_a x^{-a})$ that $s(a) - s(x) + s(x +_a y) = s(y) + s(a)$. Therefore, $s(x +_a y) = s(x) + s(y)$ and so s is a Riečan state.

Conversely, suppose that s is a Riečan state on A. Let $x, y \in A$. Since $x \wedge y \leq x$, for any $a \in Id(A)$ such that $x, y \leq a$, we have $(x \wedge y)^{-a-a} \leq x^{-a-a}$, i.e. $x^{-a} \perp (x \wedge y)$. Then $s(x^{-a} +_a (x \wedge y)) = s(x^{-a}) + s(x \wedge y) = s(a) - s(x) + s(x \wedge y)$. Moreover, it follows from Proposition 2.2(vii), (x) and Proposition 3.11(iii) that $s(x^{-a} +_a (x \wedge y)) = s(x^{-a-a} \to_a (x \wedge y)^{-a-a}) = s((x \to_a (x \wedge y))^{-a-a}) = s(x \to_a (x \wedge y)) = s(x \to_a y)$, which entails $s(x \to_a y) = s(a) - s(x) + s(x \wedge y)$. By Theorem 3.6, s is a Bosbach state. $\qquad\square$

Use $[0,1]_{\text{Ł}}$ to denote the hoop $([0,1], \odot, \to, 1)$, where $x \odot y = (x + y - 1) \vee 0$ and $x \to y = (y - x + 1) \wedge 1$ for any $x, y \in [0,1]$.

Definition 3.13. *Let A be an Ehoop with a bottom element 0. A state-morphism on A is a mapping $s : A \longrightarrow [0,1]_{\text{Ł}}$ satisfying the following conditions:*

(i) *for any $x, y \in A$, $s(x \wedge y) = s(x) \wedge s(y)$;*

(ii) *for any $x, y \in A$, $s(x \odot y) = s(x) \odot s(y)$;*

(iii) *for any $b \in Id(A)$, $x, y \in A_b = \{z \in A | z \leq b\}$, $s(x \rightarrow_b y) = (s(x) \rightarrow s(y)) \wedge s(b)$;*

(iv) *$s(0) = 0$ and there exists $x_0 \in A$ such that $s(x_0) = 1$.*

We note that a state-morphism s on an Ehoop A with a bottom element 0 is an Ehoop homomorphism from A into $[0, 1]_{\text{Ł}}$ which satisfies that there exists $x_0 \in A$ such that $s(x_0) = 1$ and $s(0) = 0$.

Example 3.14. *Let $\mathfrak{M}$ be the set of all finite subsets of $\mathbb{N}_+$, where $\mathbb{N}_+$ is the set of positive integers. Set $\wedge = \odot = \cap$ on $\mathfrak{M}$. By [27, Example 3.5], $\mathfrak{M}$ is an Ehoop. For any positive integer n, we define $s_n : \mathfrak{M} \longrightarrow [0, 1]_{\text{Ł}}$ by*

$$s_n(X) = \begin{cases} 1, & \text{if } n \in X, \\ 0, & \text{if } n \notin X. \end{cases}$$

We can check that $s_1, s_2, s_3, \cdots$ are state-morphisms on $\mathfrak{M}$.

Example 3.15. *Put $\mathfrak{X} = A_1 \times \mathfrak{M} \times \mathfrak{M} \times \cdots \times \mathfrak{M}$, where A_1 is the hoop from Example 3.4 and $\mathfrak{M}$ is the Ehoop from Example 3.14. Then $\mathfrak{X}$ is an Ehoop. Define a mapping $s : A_1 \longrightarrow [0, 1]_{\text{Ł}}$ as follows:*

$$s(x) = \begin{cases} 0, & x = 0, c, \\ \frac{1}{2}, & x = b, d, \\ 1, & x = a, 1. \end{cases}$$

Define a mapping $h : \mathfrak{X} \longrightarrow [0, 1]_{\text{Ł}}$ by $h(x_1, x_2, x_3, \cdots, x_n, \cdots) = (s(x_1), s_1(x_2), s_2(x_3), \cdots, s_{n-1}(x_n), \cdots)$, where $x_1 \in A_1$, $x_2, x_3, \cdots, x_n, \cdots \in \mathfrak{M}$ and $s_1, s_2, \cdots, s_n, \cdots$ are mappings from Example 3.14. Then h is a state-morphism on $\mathfrak{X}$.

Proposition 3.16. *Let s be a state-morphism on an Ehoop A with a bottom element 0. Then*

(i) *for any $x \in A$, $a \in Id(A)$ with $x \leq a$, we have $s(x) \leq s(a)$ and $s(x^{-a}) = s(a) - s(x)$;*

(ii) *for any $x \in A$, $a \in Id(A)$ with $x \leq a$, we have $s(x^{-a-a}) = s(x)$;*

(iii) *for any $a \in Id(A)$, $s(a) = \{0, 1\}$;*

(iv) *for any $x, y \in A$, $x \leq y$ implies $s(x) \leq s(y)$.*

Proof. (i) Let $x \in A$ and $a \in Id(A)$ such that $x \leq a$. Since $s(a) = s(a \odot a) = s(a) \odot s(a)$, we obtain $s(a) \in Id([0,1]_{\text{Ł}})$, which yields that

$$
\begin{aligned}
s(x^{-a}) &= (s(x) \to s(0)) \wedge s(a) \\
&= (s(x) \to 0) \odot s(a) \quad (\text{by } s(a) \in Id([0,1]_{\text{Ł}})) \\
&= (-s(x) + 1) \odot s(a) \\
&= (-s(x) + 1 + s(a) - 1) \vee 0 \\
&= (s(a) - s(x)) \vee 0.
\end{aligned}
$$

Moreover, we have $s(x) = s(x \wedge a) = s(x) \wedge s(a)$, so that $s(x) \leq s(a)$. This implies that $s(x^{-a}) = s(a) - s(x)$.

(ii) It is obvious by (i).

(iii) Let $a \in Id(A)$. Then $s(a) \in Id([0,1]_{\text{Ł}})$ and $s(a) = s(a) \odot s(a) = (s(a) + s(a) - 1) \vee 0$. If $s(a) \neq 0$, then $(s(a) + s(a) - 1) \vee 0 > 0$, which means $s(a) = s(a) \odot s(a) = s(a) + s(a) - 1 > 0$. Thus, $s(a) = 1$.

(iv) For any $x, y \in A$, there exists $a \in Id(A)$ such that $x, y \leq a$. Then $s(a) = \{0, 1\}$. As $x \leq y$, $x \to_a y = a$. It follows that $s(a) = s(x \to_a y) = (s(x) \to s(y)) \wedge s(a)$, i.e. $s(a) \leq s(x) \to s(y)$. If $s(a) = 0$, the property (i) entails $s(x) = s(y) = 0$. If $s(a) = 1$, we have $s(x) \to s(y) = s(a) = 1$, i.e. $s(x) \leq s(y)$, which completes the proof. $\qquad\square$

Proposition 3.17. *Let s be a state-morphism on an Ehoop A with a bottom element 0. Then s is a Bosbach state and Riečan state on A.*

Proof. Let $x, y \in A$. If $x \perp y$, for any $a \in Id(A)$ with $x, y \leq a$, we have $x^{-a-a} \leq y^{-a}$. It follows from Proposition 3.16 that

$$
\begin{aligned}
s(x +_a y) &= s(x^{-a} \to_a y^{-a-a}) \\
&= (s(x^{-a}) \to s(y)) \wedge s(a) \\
&= (s(y) - s(x^{-a}) + 1) \wedge 1 \wedge s(a) \\
&= (s(y) - (s(a) - s(x)) + 1) \wedge s(a) \\
&= (s(y) - s(a) + s(x) + 1) \odot s(a) \quad (\text{by } s(a) \in Id([0,1]_{\text{Ł}})) \\
&= (s(y) - s(a) + s(x) + 1 + s(a) - 1) \vee 0 \\
&= s(x) + s(y).
\end{aligned}
$$

Therefore, s is a Riečan state on A. By Theorem 3.12, s is also a Bosbach state on A. $\qquad\square$

Proposition 3.18. *Let s be a state-morphism on an Ehoop A with a bottom element 0. Then $Ker(s) = \{x \in A | s(x) = 1\}$ is a maximal filter of A.*

Proof. Suppose $x, y \in A$ such that $x \leq y$ and $x \in Ker(s)$. It follows from Proposition 3.16(iv) that $s(y) = 1$, i.e. $y \in Ker(s)$. Let $x, y \in Ker(s)$. Then $s(x \odot y) = s(x) \odot s(y) = 1 \odot 1 = 1$. That is, $x \odot y \in Ker(s)$. Since there exists $x_0 \in A$ such that $s(x_0) = 1$, we have $a \in Id(A)$ such that $x_0, x \leq a$ for any $x \in A$. It follows that $s(x_0) \leq s(a)$ and so $s(a) = 1$, i.e. $a \in Ker(s)$. Therefore, for any $x \in A$, there exists $a \in Id(A) \cap Ker(s)$ such that $x \leq a$. Hence, $Ker(s)$ is a filter of A.

As $s(0) = 0$, $Ker(s) \neq A$. Suppose $x \notin Ker(s)$, i.e. $s(x) \neq 1$. For any $u \in A$, we have $s(u)^n = (ns(u) - (n-1)) \vee 0$ by mathematical induction, where $n \in \mathbb{N} \backslash \{0\}$. Let $s(x) = \frac{q}{p} < 1$, where p, q are positive integers and $\frac{q}{p}$ is irreducible. Then $q \leq p - 1$ and $s(x)^p = (ps(x) - (p-1)) \vee 0 = (q - p + 1) \vee 0 = 0$. That is to say, when $s(x) \neq 1$, there exists $n \in \mathbb{N} \backslash \{0\}$ such that $s(x)^n = 0$. It follows from Proposition 3.16(i) that $s(x^n \to_b 0) = s(b) - s(x)^n = s(b)$ for any $b \in Id(A)$ such that $x \leq b$. Since there is $x_0 \in A$ such that $s(x_0) = 1$, we have $c \in Id(A)$ such that $x, x_0 \leq c$. Then $s(x_0) \leq s(c)$ and so $s(c) = 1$. Therefore, there exist $n \in \mathbb{N} \backslash \{0\}$ and $c \in Id(A)$ with $x \leq c$ such that $s(x^n \to_c 0) = s(c) = 1$, i.e. $x^n \to_c 0 \in Ker(s)$. By Remark 2.7, $Ker(s)$ is a maximal filter of A. $\square$

Proposition 3.19. *If F is a maximal filter of an Ehoop A, then A/F is a simple hoop. In particular, if the hoop A/F is bounded, it can be viewed as a linear MV-algebra.*

Proof. Let F be a maximal filter of an Ehoop A. For any $x \in A$, there exists $a \in Id(A) \cap F$ such that $x \leq a$. Then $x/F \leq a/F = F$, which means that F is the top element of the Ehoop A/F. Hence, A/F is a hoop. Let $\mathcal{H}$ be a proper filter of A/F. Then $\mathcal{H} \neq A/F$ and $F \in \mathcal{H}$. Set $\hat{F} = \{x \in A | x/F \in \mathcal{H}\}$. Obviously, $F \subseteq \hat{F}$ and $\hat{F}$ is a filter of A. Suppose that there exists $x \in \hat{F} \backslash F$, i.e. $x/F \in \mathcal{H}$ and $x \notin F$. Choose $y \in A$ such that $y/F \notin \mathcal{H}$. By Proposition 2.6, there exist $n \in \mathbb{N} \backslash \{0\}$ and $b \in Id(A)$ with $x, y \leq b$ such that $x^n \to_b y \in F$. This together with $x^n \to_b y \leq b$ implies $b \in F$. Therefore, $(x^n \to_b y)/F = F = b/F$. It follows that $x^n/F \to_{b/F} y/F = (x^n \to_b y)/F = b/F$ and so $x^n/F \leq_{b/F} y/F$ in the hoop $(A/F)_{b/F} = \{x \in A/F | x \leq b/F\}$. That is, $x^n/F \leq y/F$. By $y/F \notin \mathcal{H}$, we have $x^n/F \notin \mathcal{H}$, i.e. $x^n \notin \hat{F}$, which is a contradiction. Thus, $\hat{F} = \{x \in A | x/F \in \mathcal{H}\} = F$ and so $\mathcal{H} = \{F\}$. This proves that the hoop A/F has only two filters: $\{F\}$ and A/F. Therefore, A/F is a simple hoop.

If A/F is bounded, A/F is a linear Wajsberg hoop and can be embedded into $[0, 1]$ by [13, Theorem 2.1]. Due to [10, Remark 2.2], each bounded Wajsberg hoop satisfies the double negation property. By [19, Proposition 4.9, 4.10], any bounded Wajsberg hoop can be viewed as an MV-algebra. This proves that when A/F is bounded, it is a linear MV-algebra. $\square$

Proposition 3.20. *Let A be an Ehoop with a bottom element 0 and s_1, s_2 be two state-morphisms on A. Then $Ker(s_1) = Ker(s_2)$ if and only if $s_1 = s_2$.*

Proof. If $s_1 = s_2$, it is obvious that $Ker(s_1) = Ker(s_2)$. Conversely, suppose $Ker(s_1) = Ker(s_2)$. Since $A/Ker(s_1)$ and $A/Ker(s_2)$ are bounded, they can be viewed as linear MV-algebras by Proposition 3.19. It follows from Proposition 3.17 that s_1 and s_2 are Bosbach states on A. Define $\tilde{s}_1$ and $\tilde{s}_2$ according to Proposition 3.7(iv). It can be known that $\tilde{s}_1$ and $\tilde{s}_2$ are state-morphisms. Set $A_i = \tilde{s}_i(A/Ker(s_i))$ for $i = 1, 2$. Then A_1, A_2 are MV-subalgebras of the standard MV-algebra. Similar to the proof of [12, Proposition 4.5], we can also get that $s_1 = s_2$. □

Proposition 3.21. *Let F be a maximal filter of an Ehoop A with a bottom element 0. Then there is a unique state-morphism s on A such that $Ker(s) = F$.*

Proof. By Proposition 3.19 and its proof, we can assume that $A/F = [0,1]_{\text{Ł}}$. Define the mapping $s : A \longrightarrow A/F$ by $s(x) = x/F$ for any $x \in A$. Clearly, s is an Ehoop homomorphism. For any $x \in F$, since F is the top element of A/F, we have $s(x) = 1$. Therefore, the mapping s is a state-morphism such that $Ker(s) = F$.

If there exists another state-morphism s' on A such that $Ker(s') = F$. Then $Ker(s) = Ker(s')$. By Proposition 3.20, we have $s = s'$, which proves the uniqueness of s. □

Due to [6], a hoop does not necessarily have a maximal filter. Therefore, a maximal filter does not always exist in each Ehoop. According to the notion of strong units of pseudo-hoops in [6], we give the notion of strong units of Ehoops. An element u of the Ehoop A is said to be a strong unit of A if the filter of A generated by u is equal to A. If the Ehoop A has a bottom element 0, we can say that 0 is a strong unit of A. A value V of $g \in A$ is a maximal element of the set $\{F|F \text{ is a filter and } g \notin F\}$.

Lemma 3.22. *If the Ehoop A has a strong unit, then there is at least one maximal filter in A.*

Proof. Suppose that u is a strong unit of A. Set $\mathcal{F} = \{F|F \text{ is a filter of } A \text{ and } u \notin F\}$. If $\{F_i|i \in I\}$ is a chain in $\mathcal{F}$, it can be easily shown that $\cup\{F_i|i \in I\}$ is a filter of A. We claim that $u \notin \cup\{F_i|i \in I\}$. Otherwise, there exists $i \in I$ such that $u \in F_i$, which is a contradiction. Thus, $\cup\{F_i|i \in I\} \in \mathcal{F}$. By Zorn's Lemma, there exists a maximal element V in $\mathcal{F}$. Hence, V is a value of u. If there is a filter G of A such that V is strictly contained in G, then $u \in G$ and so the filter of A generated by u is contained in G. Since u is a strong unit, we have $G = A$, which means that V is a maximal filter of A. □

Theorem 3.23. *Each Ehoop with a bottom element 0 admits a Bosbach state and a Riečan state.*

Proof. Let A be an Ehoop with a bottom element 0. It is obvious that 0 is a strong unit. By Lemma 3.22, there is at least a maximal filter F of A. It follows from Proposition 3.21 that there is a state-morphism s on A. By Proposition 3.17, s is a Bosbach state and a Riečan state. $\qquad\square$

4 Internal states on Ehoops

In this section, we introduce the notion of internal states on Ehoops with a bottom element. Prime state ideal theorem on state Ehoops is given. Also, it is proved that if a state Ehoop (A, τ) satisfies the double negation property, the set of all prime state ideals of (A, τ) is a topological space.

Definition 4.1. *Let A be an Ehoop with a bottom element 0. An internal state on A is a mapping $\tau : A \longrightarrow A$ which satisfies the following conditions:*

(IS1) $\tau(0) = 0$;
(IS2) for any $x, y \in A$, for any $a \in Id(A)$ with $x, y \leq a$, $\tau(x \to_a y) = \tau(x) \to_{\tau(a)} \tau(x \wedge y)$;
(IS3) for any $x, y \in A$, for any $a \in Id(A)$ with $x, y \leq a$, $\tau(x \odot y) = \tau(x) \odot \tau(x \to_a (x \odot y))$;
(IS4) for any $x, y \in A$, $\tau(\tau(x) \odot \tau(y)) = \tau(x) \odot \tau(y)$;
(IS5) for any $x, y \in A$, $\tau(\tau(x) \wedge \tau(y)) = \tau(x) \wedge \tau(y)$.

The pair (A, τ) is called a state Ehoop.

Example 4.2. *Let A, B be two Ehoops with a bottom element. Define $\tau : A \times B \longrightarrow A \times B$ by $\tau(x, y) = (x, 0)$ for any $(x, y) \in A \times B$. Then τ is an internal state on $A \times B$.*

Example 4.3. *Let A be the Ehoop from Example 3.3. Define a mapping $\tau : A \longrightarrow A$ by*

$$\tau(u) = \begin{cases} u, & \text{if } u \in [0, 1], \\ 2, & \text{if } u > 1. \end{cases}$$

We can check that τ is an internal state on A.

Example 4.4. *Let* $\mathcal{M}$ *be the Ehoop from Example 3.14. Define a mapping* $\tau : \mathcal{M} \longrightarrow \mathcal{M}$ *as follows:*

$$\tau(X) = \begin{cases} \{3,5\}, & if\ 3,5 \in X, \\ \{3\}, & if\ 3 \in X,\ 5 \notin X, \\ \{5\}, & if\ 5 \in X,\ 3 \notin X, \\ \emptyset, & if\ X = \emptyset\ or\ 3,5 \notin X. \end{cases}$$

Then τ *is an internal state on* $\mathcal{M}$.

Next, we give some properties of internal states on Ehoops with a bottom element.

Proposition 4.5. *Let* τ *be an internal state on an Ehoop* A *with a bottom element* 0. *Then the following properties hold:*

(i) *for any* $a \in Id(A)$, $\tau(a) \in Id(A)$;

(ii) *for any* $x,y \in A$, *if* $x \leq y$, *then* $\tau(x) \leq \tau(y)$;

(iii) *for any* $x \in A$, *for any* $a \in Id(A)$ *with* $x \leq a$, $\tau(x^{-a}) = \tau(x) \rightarrow_{\tau(a)} 0$;

(iv) *for any* $x,y \in A$, $\tau(x \odot y) \geq \tau(x) \odot \tau(y)$; *if* $x \odot y = 0$, *then* $\tau(x \odot y) = \tau(x) \odot \tau(y)$;

(v) *for any* $x,y \in A$, *for any* $a \in Id(A)$ *with* $x,y \leq a$, *we have* $\tau(x \rightarrow_a y) \leq \tau(x) \rightarrow_{\tau(a)} \tau(y)$. *If* x *and* y *are comparable, then* $\tau(x \rightarrow_a y) = \tau(x) \rightarrow_{\tau(a)} \tau(y)$;

(vi) *for any* $x \in A$, $\tau^2(x) = \tau(x)$;

(vii) *for any* $x,y \in A$, *if* $x \perp y$, *then* $\tau(x) \perp \tau(y)$;

(viii) *for any* $x,y \in A$, *if* $x \perp y$, *then* $\tau(\tau(x) +_{\tau(a)} \tau(y)) = \tau(x) +_{\tau(a)} \tau(y)$, *where* a *is an arbitrary idempotent element of* A *such that* $x,y \leq a$;

(ix) $\tau(A) = \{x \in A | \tau(x) = x\}$;

(x) *for any* $x,y \in A$, *for any* $a \in Id(A)$ *such that* $x,y \leq a$, $\tau(x) \odot \tau(x \rightarrow_a y) = \tau(y) \odot \tau(y \rightarrow_a x) = \tau(x \wedge y)$;

(xi) *for any* $x,y \in A$, *for any* $a \in Id(A)$ *with* $x,y \leq a$, $\tau(x \ominus_a y) \leq \tau(x) \ominus_{\tau(a)} \tau(y)$. *If* x^{-a} *and* y *are comparable, then* $\tau(x \ominus_a y) = \tau(x) \ominus_{\tau(a)} \tau(y)$.

Proof. (i) Let $a \in Id(A)$. The condition $(IS3)$ implies $\tau(a) = \tau(a \odot a) = \tau(a) \odot \tau(a \rightarrow_a (a \odot a)) = \tau(a) \odot \tau(a)$. That is, $\tau(a) \in Id(A)$.

(ii) Let $x \leq y$. For any $a \in Id(A)$ with $x,y \leq a$, we have $x = x \wedge y = y \odot (y \rightarrow_a x)$. It follows from $(IS3)$ that $\tau(x) = \tau(y \odot (y \rightarrow_a x)) = \tau(y) \odot \tau(y \rightarrow_a (y \odot (y \rightarrow_a x))) \leq \tau(y)$.

(iii) By $(IS1)$ and $(IS2)$, we have $\tau(x^{-a}) = \tau(x) \rightarrow_{\tau(a)} \tau(x \wedge 0) = \tau(x) \rightarrow_{\tau(a)} 0$.

(iv) Let $x,y \in A$ and $a \in Id(A)$ such that $x,y \leq a$. As $y \leq x \rightarrow_a (x \odot y)$, we have $\tau(y) \leq \tau(x \rightarrow_a (x \odot y))$, which yields $\tau(x \odot y) = \tau(x) \odot \tau(x \rightarrow_a (x \odot y)) \geq \tau(x) \odot \tau(y)$ by $(IS3)$. If $x \odot y = 0$, then $\tau(x \odot y) = 0$. Since $0 = \tau(x \odot y) \geq \tau(x) \odot \tau(y)$, we have $\tau(x \odot y) = \tau(x) \odot \tau(y)$.

(v) Suppose $x, y \in A$ and $a \in Id(A)$ such that $x, y \leq a$. Then $\tau(x \to_a y) = \tau(x) \to_{\tau(a)} \tau(x \wedge y)$. From $\tau(x \wedge y) \leq \tau(y)$ it follows that $\tau(x \to_a y) \leq \tau(x) \to_{\tau(a)} \tau(y)$. If $x \leq y$, then $\tau(x) \leq \tau(y)$ and so $\tau(x) \to_{\tau(a)} \tau(y) = \tau(a)$. This together with $\tau(x \to_a y) = \tau(x) \to_{\tau(a)} \tau(x \wedge y) = \tau(x) \to_{\tau(a)} \tau(x) = \tau(a)$ implies $\tau(x \to_a y) = \tau(x) \to_{\tau(a)} \tau(y)$. If $y \leq x$, then $\tau(x \to_a y) = \tau(x) \to_{\tau(a)} \tau(x \wedge y) = \tau(x) \to_{\tau(a)} \tau(y)$, which completes the proof.

(vi) Let $x \in A$ and $a \in Id(A)$ such that $x \leq a$. Applying $(IS5)$, we have $\tau^2(x) = \tau(\tau(x)) = \tau(\tau(x) \wedge \tau(a)) = \tau(x) \wedge \tau(a) = \tau(x)$.

(vii) Suppose $x \perp y$. For any $a \in Id(A)$ such that $x, y \leq a$, we have $y^{-a-a} \leq x^{-a}$ and so $\tau(y^{-a-a}) \leq \tau(x^{-a})$. That is, $(\tau(y) \to_{\tau(a)} 0) \to_{\tau(a)} 0 \leq \tau(x) \to_{\tau(a)} 0$. As $\tau(x), \tau(y) \leq \tau(a) \in Id(A)$, $\tau(x) \perp \tau(y)$.

(viii) Let $x \perp y$. Then $\tau(x) \perp \tau(y)$. For all $a \in Id(A)$ with $x, y \leq a$, we have $\tau(y^{-a-a}) \leq \tau(x^{-a})$. It follows that

$$
\begin{aligned}
\tau(\tau(x) +_{\tau(a)} \tau(y)) &= \tau(\tau(x)^{-\tau(a)} \to_{\tau(a)} \tau(y)^{-\tau(a)-\tau(a)}) \\
&= \tau(\tau(x^{-a}) \to_{\tau(a)} \tau(y^{-a-a})) \quad \text{(by (iii))} \\
&= \tau^2(x^{-a}) \to_{\tau(a)} \tau(\tau(x^{-a}) \wedge \tau(y^{-a-a})) \quad \text{(by (IS2))} \\
&= \tau(x^{-a}) \to_{\tau(a)} \tau^2(y^{-a-a}) \quad \text{(by } \tau(y^{-a-a}) \leq \tau(x^{-a})) \\
&= \tau(x^{-a}) \to_{\tau(a)} \tau(y^{-a-a}) \\
&= \tau(x) +_{\tau(a)} \tau(y).
\end{aligned}
$$

(ix) Clearly, $\{x \in A | \tau(x) = x\} \subseteq \tau(A)$. Conversely, suppose $y \in \tau(A)$. There exists $x \in A$ such that $\tau(x) = y$, which entails $\tau(y) = \tau(\tau(x)) = \tau(x) = y$ and so $y \in \{x \in A | \tau(x) = x\}$. This means that $\tau(A) \subseteq \{x \in A | \tau(x) = x\}$. Therefore, $\tau(A) = \{x \in A | \tau(x) = x\}$.

(x) Let $x, y \in A$ and $a \in Id(A)$ such that $x, y \leq a$. The condition $(IS3)$ implies

$$
\begin{aligned}
\tau(x \wedge y) &= \tau(x \odot (x \to_a y)) \\
&= \tau(x) \odot \tau(x \to_a (x \odot (x \to_a y))) \\
&= \tau(x) \odot \tau(x \to_a (x \wedge y)) \\
&= \tau(x) \odot \tau(x \to_a y).
\end{aligned}
$$

In a similar way, we can show $\tau(x \wedge y) = \tau(y) \odot \tau(y \to_a x)$.

(xi) Let $x, y \in A$ and $a \in Id(A)$ such that $x, y \leq a$. By (iii) and (v), we have $\tau(x \ominus_a y) = \tau(x^{-a} \to_a y) \leq \tau(x^{-a}) \to_{\tau(a)} \tau(y) = (\tau(x) \to_{\tau(a)} 0) \to_{\tau(a)} \tau(y) = \tau(x) \ominus_{\tau(a)} \tau(y)$. If x^{-a} and y are comparable, we obtain $\tau(x \ominus_a y) = \tau(x) \ominus_{\tau(a)} \tau(y)$ by (v). $\qquad \square$

Lemma 4.6. *Let A be an Ehoop with a bottom element 0 and τ an internal state on A. Then $\tau(A)$ is a subalgebra of A.*

Proof. By $(IS4)$ and $(IS5)$, $\tau(A)$ is closed under $\odot$ and $\wedge$. For any $b \in Id(A) \cap \tau(A)$, we shall show that $\tau(A)$ is closed under $\to_b$. Suppose $u, v \in \tau(A)$ and $u, v \leq b$. There exist $x, y \in A$ such that $u = \tau(x) \leq b$ and $v = \tau(y) \leq b$. Since $b \in \tau(A)$, we have $\tau(b) = b$ by Proposition 4.5(ix). It follows that

$$
\begin{aligned}
\tau(u \to_b v) &= \tau(\tau(x) \to_b \tau(y)) \\
&= \tau(\tau(x)) \to_{\tau(b)} \tau(\tau(x) \wedge \tau(y)) \quad \text{(by (IS2))} \\
&= \tau(x) \to_{\tau(b)} (\tau(x) \wedge \tau(y)) \quad \text{(by (IS5))} \\
&= u \to_b (u \wedge v) \\
&= u \to_b v
\end{aligned}
$$

and so $u \to_b v \in \tau(A)$. Hence, it can be easily known that $\tau(A)_b = \{x \in \tau(A) | x \leq b\}$ is a hoop. Moreover, for any $u, v \in \tau(A)$, there exist $x, y \in A$ such that $\tau(x) = u$ and $\tau(y) = v$. Set $a \in Id(A)$ such that $x, y \leq a$. Then $\tau(a) \in Id(A) \cap \tau(A)$ such that $u = \tau(x) \leq \tau(a)$ and $v = \tau(y) \leq \tau(a)$. Therefore, $\tau(A)$ is a subalgebra of A. $\qquad\square$

Let A be an Ehoop with a bottom element 0, τ an internal state on A and s a Riečan state on A. If $\tau(x) = \tau(y)$ implies $s(x) = s(y)$ for any $x, y \in A$, the mapping s is called τ-compatible. The set of all τ-compatible Riečan states on A is denoted by $S_\tau[A]$.

Theorem 4.7. *Let (A, τ) be a state Ehoop. Then there is a one-to-one correspondence between the set of all τ-compatible Riečan states on A and the set of all Riečan states on $\tau(A)$.*

Proof. Denote the set of all Riečan states on $\tau(A)$ by $S[\tau(A)]$. Suppose $s \in S[\tau(A)]$. The mapping $\varphi : S[\tau(A)] \longrightarrow S_\tau[A]$ is defined by $\varphi(s)(x) = s(\tau(x))$ for any $x \in A$. Since s is a Riečan state on $\tau(A)$, there exists $x_0 \in \tau(A)$ such that $s(x_0) = 1$. Set $x_0 = \tau(y_0)$, where $y_0 \in A$. Then $\varphi(s)(y_0) = s(\tau(y_0)) = s(x_0) = 1$. Suppose $x, y \in A$ such that $x \perp y$. Then $\tau(x) \perp \tau(y)$. For any $b \in Id(A)$ such that $x, y \leq b$, we have $\tau(x +_b y) = \tau(x^{-b} \to_b y^{-b-b}) = \tau(x^{-b}) \to_{\tau(b)} \tau(x^{-b} \wedge y^{-b-b}) = \tau(x^{-b}) \to_{\tau(b)} \tau(y^{-b-b}) = \tau(x) +_{\tau(b)} \tau(y)$, which yields

$$
\begin{aligned}
\varphi(s)(x +_b y) &= s(\tau(x +_b y)) \\
&= s(\tau(x) +_{\tau(b)} \tau(y)) \\
&= s(\tau(x)) + s(\tau(y)) \quad (s \text{ is a Riečan state}) \\
&= \varphi(s)(x) + \varphi(s)(y).
\end{aligned}
$$

Moreover, let $x, y \in A$ and $\tau(x) = \tau(y)$. Then $\varphi(s)(x) = s(\tau(x)) = s(\tau(y)) = \varphi(s)(y)$. Therefore, $\varphi(s)$ is a τ-compatible Riečan state on A.

Suppose that m is a τ-compatible Riečan state on A. Define $\psi : S_\tau[A] \longrightarrow S[\tau(A)]$ by $\psi(m)(\tau(x)) = m(x)$ for any $x \in A$. We shall show that $\psi(m)$ is a Riečan state on $\tau(A)$. Suppose $x, y \in A$ and $\tau(x) \perp \tau(y)$. Let $a \in Id(A) \cap \tau(A)$ with $\tau(x), \tau(y) \leq a$. There exists $b \in A$ such that $\tau(b) = a$. It follows from Proposition 4.5(viii) that $\tau(\tau(x) +_a \tau(y)) = \tau(\tau^2(x) +_{\tau(a)} \tau^2(y)) = \tau(\tau^2(x)) +_{\tau(a)} \tau(\tau^2(y)) = \tau(x) +_a \tau(y)$ and so

$$
\begin{aligned}
\psi(m)(\tau(x) +_a \tau(y)) &= \psi(m)(\tau(\tau(x) +_a \tau(y))) \\
&= m(\tau(x) +_a \tau(y)) \\
&= m(\tau(x)) + m(\tau(y)) \\
&= \psi(m)(\tau(\tau(x))) + \psi(m)(\tau(\tau(y))) \\
&= \psi(m)(\tau(x)) + \psi(m)(\tau(y)).
\end{aligned}
$$

Since m is a Riečan state on A, there exists $x_0 \in A$ such that $m(x_0) = 1$. Then $\psi(m)(\tau(x_0)) = m(x_0) = 1$. Hence, $\psi(m)$ is a Riečan state on $\tau(A)$.

Suppose $m_1, m_2 \in S_\tau[A]$ such that $\psi(m_1) = \psi(m_2)$. For any $x \in A$, we have $\psi(m_1)(\tau(x)) = \psi(m_2)(\tau(x))$, i.e. $m_1(x) = m_2(x)$. Hence, $m_1 = m_2$ and so ψ is injective. Suppose that s is a Riečan state on $\tau(A)$. Then $\psi(\varphi(s))(\tau(x)) = \varphi(s)(x) = s(\tau(x))$ for any $x \in A$, i.e. $\psi(\varphi(s)) = s$. Therefore, ψ is surjective. We prove that ψ is bijective. $\qquad\square$

In what follows, we introduce state ideals of state Ehoops.

Definition 4.8. *Let (A, τ) be a state Ehoop. If I is an ideal of A satisfying that $x \in I$ implies $\tau(x) \in I$ for any $x \in A$, I is called a state ideal of (A, τ).*

Example 4.9. *Let $\mathfrak{M}$ be the Ehoop from Example 3.14 and τ the mapping from Example 4.4. By [27, Example 4.5], for any $\emptyset \neq M \in \mathfrak{M}$, the set of all subsets of M is an ideal of $\mathfrak{M}$. Clearly, it is also a state ideal of $(\mathfrak{M}, \tau)$.*

Let (A, τ) be a state Ehoop. Clearly, the intersection of all state ideals of (A, τ) is still a state ideal. Denote by $SI(A)$ the set of all state ideals of (A, τ). If $I \in SI(A)$ and $I \neq A$, I is said to be proper. For any subset X of A, the state ideal of (A, τ) generated by X is denoted by $< X >_\tau$. If $X = \{x\}$, we use $< x >_\tau$ instead of $< \{x\} >_\tau$.

Proposition 4.10. *Let (A, τ) be a state Ehoop and $x \in A$. For any $a \in Id(A)$ with $x, \tau(x) \leq a$, we have $< x >_\tau = < x \ominus_a \tau(x) >_\tau$.*

Proof. It is clear that $x \in < x >_\tau$ and $\tau(x) \in < x >_\tau$, which yield $x \ominus_a \tau(x) \in < x >_\tau$. Then $< x \ominus_a \tau(x) >_\tau \subseteq < x >_\tau$. Since $x \leq x \ominus_a \tau(x)$, we obtain $x \in < x \ominus_a \tau(x) >_\tau$ and so $< x >_\tau \subseteq < x \ominus_a \tau(x) >_\tau$. Thus, $< x >_\tau = < x \ominus_a \tau(x) >_\tau$. □

Proposition 4.11. *Let (A, τ) be a state Ehoop, $x \in A$, $X \subseteq A$ and I a state ideal of (A, τ). If A is normal, we have*

    (i) $< x >_\tau = \{y \in A \mid y \leq n_a(x \ominus_a \tau(x)),$ *for some* $n \in \mathbb{N} \backslash \{0\}$, $a \in Id(A)$ *with* $x,$
        $\tau(x) \leq a\}$;

    (ii) $< X >_\tau = \{y \in A \mid y \leq x_1 \ominus_a \tau(x_1) \ominus_a x_2 \ominus_a \tau(x_2) \ominus_a \cdots \ominus_a x_n \ominus_a \tau(x_n),$ *for some*
        $n \in \mathbb{N} \backslash \{0\}$, $x_1, x_2, \cdots, x_n \in X$, $a \in Id(A)$ *with* $x_1, \tau(x_1), x_2, \tau(x_2), \cdots, x_n, \tau($
        $x_n) \leq a\}$;

    (iii) $< I \cup \{x\} >_\tau = \{y \in A | y \leq \ominus_{a \, i=1}^{l}(w_i \ominus_a n_{ia}(x \ominus_a \tau(x))),$ *for some* $w_i \in I$, $l, n_i \in$
        $\mathbb{N} \backslash \{0\}$, *and* $a \in Id(A)$ *with* $w_i, x, \tau(x) \leq a\}$.

Proof. The proofs are obvious. □

Proposition 4.12. *Let A be an Ehoop with the double negation property and τ an internal state on A. For any $x, y \in A$, there exists $a \in Id(A)$ such that $x, \tau(x), y, \tau(y) \leq a$ and $< x >_\tau \cap < y >_\tau \subseteq < (x \ominus_a \tau(x)) \wedge (y \ominus_a \tau(y)) >_\tau$.*

Proof. Let $h \in < x >_\tau \cap < y >_\tau$. There exist $s, t \in \mathbb{N} \backslash \{0\}$, $a, b \in Id(A)$ with $x, \tau(x) \leq a$ and $y, \tau(y) \leq b$ such that $h \leq s_a(x \ominus_a \tau(x))$ and $h \leq t_b(y \ominus_b \tau(y))$. Choose $c \in Id(A)$ such that $a, b \leq c$. From Proposition 2.4(ii) and Proposition 2.5(vi) it follows that

$$h \leq s_a(x \ominus_a \tau(x)) \wedge t_b(y \ominus_b \tau(y))$$
$$\leq s_c(x \ominus_c \tau(x)) \wedge t_c(y \ominus_c \tau(y))$$
$$\leq (ts)_c((x \ominus_c \tau(x)) \wedge (y \ominus_c \tau(y)))$$

and so $h \in < (x \ominus_c \tau(x)) \wedge (y \ominus_c \tau(y)) >_\tau$. This proves that $< x >_\tau \cap < y >_\tau \subseteq < (x \ominus_c \tau(x)) \wedge (y \ominus_c \tau(y)) >_\tau$. □

Proposition 4.13. *Let (A, τ) be a state Ehoop. For any $x, y \in A$, any $a, b \in Id(A)$ with $x, \tau(x) \leq a$ and $y, \tau(y) \leq b$, we have $< (x \ominus_a \tau(x)) \wedge (y \ominus_b \tau(y)) >_\tau \subseteq < x >_\tau \cap < y >_\tau$.*

Proof. Let $x, y \in A$ and $a, b \in Id(A)$ such that $x, \tau(x) \leq a$ and $y, \tau(y) \leq b$. Then we have $(x \ominus_a \tau(x)) \wedge (y \ominus_b \tau(y)) \leq x \ominus_a \tau(x) \in < x >_\tau$ and $(x \ominus_a \tau(x)) \wedge (y \ominus_b \tau(y)) \leq y \ominus_b \tau(y) \in < y >_\tau$. Therefore, $(x \ominus_a \tau(x)) \wedge (y \ominus_b \tau(y)) \in < x >_\tau \cap < y >_\tau$. This implies $< (x \ominus_a \tau(x)) \wedge (y \ominus_b \tau(y)) >_\tau \subseteq < x >_\tau \cap < y >_\tau$. □

Proposition 4.14. *Let A be an Ehoop with the double negation property and τ an internal state on A. For any $x, y \in A$, there exists $a \in Id(A)$ such that $x, \tau(x), y, \tau(y) \leq a$ and $<x>_\tau \cap <y>_\tau = <(x \ominus_a \tau(x)) \wedge (y \ominus_a \tau(y))>_\tau$.*

Proof. By Proposition 4.12 and Proposition 4.13, the proof is straightforward. $\square$

Proposition 4.15. *Let A be an Ehoop with the double negation property, τ an internal state on A and I a state ideal of (A, τ). For any $x, y \in A$, if there exists $a \in Id(A)$ such that $x, \tau(x), y, \tau(y) \leq a$ and $(x \ominus_a \tau(x)) \wedge (y \ominus_a \tau(y)) \in I$, then for all $b \in Id(A)$ such that $x, \tau(x), y, \tau(y) \leq b$, we have $(x \ominus_b \tau(x)) \wedge (y \ominus_b \tau(y)) \in I$.*

Proof. Let $x, y \in A$. Suppose $(x \ominus_a \tau(x)) \wedge (y \ominus_a \tau(y)) \in I$, where $a \in Id(A)$ such that $x, \tau(x), y, \tau(y) \leq a$. As $x, \tau(x) \leq x \ominus_a \tau(x)$ and $y, \tau(y) \leq y \ominus_a \tau(y)$, we have $x \wedge y, \tau(x) \wedge y, x \wedge \tau(y), \tau(x) \wedge \tau(y) \leq (x \ominus_a \tau(x)) \wedge (y \ominus_a \tau(y))$ and so $x \wedge y, \tau(x) \wedge y, x \wedge \tau(y), \tau(x) \wedge \tau(y) \in I$. Let b be an arbitrary idempotent element of A such that $x, \tau(x), y, \tau(y) \leq b$. It follows that $(x \wedge y) \ominus_b (\tau(x) \wedge y) \ominus_b (x \wedge \tau(y)) \ominus_b (\tau(x) \wedge \tau(y)) \in I$. By Proposition 2.5(vi), $(x \ominus_b \tau(x)) \wedge (y \ominus_b \tau(y)) \leq (x \wedge y) \ominus_b (\tau(x) \wedge y) \ominus_b (x \wedge \tau(y)) \ominus_b (\tau(x) \wedge \tau(y))$. This together with $I \in SI(A)$ implies $(x \ominus_b \tau(x)) \wedge (y \ominus_b \tau(y)) \in I$. $\square$

Definition 4.16. *Let P be a state ideal of a state Ehoop (A, τ). If P is proper and for any $x, y \in A$, any $a \in Id(A)$ with $x, \tau(x), y, \tau(y) \leq a$, $(x \ominus_a \tau(x)) \wedge (y \ominus_a \tau(y)) \in P$ implies $x \in P$ or $y \in P$, P is said to be a prime state ideal of (A, τ).*

Let (A, τ) be a state Ehoop. Denote by $PSI(A)$ the set of all prime state ideals of (A, τ).

Proposition 4.17. *Let A be an Ehoop with the double negation property, τ an internal state on A and P a proper state ideal of (A, τ). Then $P \in PSI(A)$ if and only if for any $I_1, I_2 \in SI(A)$, $I_1 \cap I_2 \subseteq P$ implies $I_1 \subseteq P$ or $I_2 \subseteq P$.*

Proof. Let $P \in PSI(A)$. Suppose that $I_1, I_2 \in SI(A)$ such that $I_1 \cap I_2 \subseteq P$, $I_1 \nsubseteq P$ and $I_2 \nsubseteq P$. There exist $x_1, x_2 \in A$ such that $x_1 \in I_1 \backslash P$ and $x_2 \in I_2 \backslash P$. For any $a \in Id(A)$ with $x_1, \tau(x_1), x_2, \tau(x_2) \leq a$, we have $x_1 \ominus_a \tau(x_1) \in I_1$ and $x_2 \ominus_a \tau(x_2) \in I_2$. It follows that $(x_1 \ominus_a \tau(x_1)) \wedge (x_2 \ominus_a \tau(x_2)) \in I_1 \cap I_2 \subseteq P$. Since P is prime, we obtain $x_1 \in P$ or $x_2 \in P$, which is a contradiction. Therefore, for any $I_1, I_2 \in SI(A)$, if $I_1 \cap I_2 \subseteq P$, then $I_1 \subseteq P$ or $I_2 \subseteq P$.

Conversely, suppose that for any $I_1, I_2 \in SI(A)$, $I_1 \cap I_2 \subseteq P$ implies $I_1 \subseteq P$ or $I_2 \subseteq P$. Let $x, y \in A$, $a \in Id(A)$ such that $x, \tau(x), y, \tau(y) \leq a$ and $(x \ominus_a \tau(x)) \wedge (y \ominus_a \tau(y)) \in P$. By Proposition 4.14, there exists $b \in Id(A)$ such that $x, \tau(x), y, \tau(y) \leq b$ and $<x>_\tau \cap <y>_\tau = <(x \ominus_b \tau(x)) \wedge (y \ominus_b \tau(y))>_\tau$. It follows

629

from Proposition 4.15 that $(x \ominus_b \tau(x)) \wedge (y \ominus_b \tau(y)) \in P$ and so $< x >_\tau \cap < y >_\tau = <$ $(x \ominus_b \tau(x)) \wedge (y \ominus_b \tau(y)) >_\tau \subseteq P$. By the assumption, we obtain $< x >_\tau \subseteq P$ or $< y >_\tau \subseteq P$. Thus, $x \in P$ or $y \in P$. This proves that P is a prime state ideal of (A, τ). $\qquad\square$

Theorem 4.18. *(Prime state ideal theorem) Let A be an Ehoop with the double negation property and τ an internal state on A. If $I \in SI(A)$ and $x \in A \backslash I$, then there exists $P \in PSI(A)$ such that $I \subseteq P$ and $x \notin P$.*

Proof. Set $\Lambda = \{Q \in SI(A) | x \notin Q \text{ and } I \subseteq Q\}$. Then $I \in \Lambda$ and so $\Lambda \neq \emptyset$. By Zorn's Lemma, there is a maximal element P in Λ. Clearly, $P \neq A$. We claim that $P \in PSI(A)$. Indeed, let $u, v \in A$ and $a \in Id(A)$ with $u, v \leq a$. Suppose $(u \ominus_a \tau(u)) \wedge (v \ominus_a \tau(v)) \in P$ but $u, v \notin P$. It is obvious that $< P \cup \{u\} >_\tau \notin \Lambda$ and $< P \cup \{v\} >_\tau \notin \Lambda$. Thus, $x \in < P \cup \{u\} >_\tau$ and $x \in < P \cup \{v\} >_\tau$. Then we obtain $x \leq \ominus_{b i=1}^s (p_i \ominus_b n_{ib}(u \ominus_b \tau(u)))$, for some $p_i \in I$, $s, n_i \in \mathbb{N} \backslash \{0\}$, and $b \in Id(A)$ with $p_i, u, \tau(u) \leq b$. Also, we have $x \leq \ominus_{c i=1}^t (q_i \ominus_c m_{ic}(v \ominus_c \tau(v)))$, for some $q_i \in I$, $t, m_i \in \mathbb{N} \backslash \{0\}$, and $c \in Id(A)$ with $q_i, v, \tau(v) \leq c$. Choose $d \in Id(A)$ such that $b, c \leq d$. Set $r = (\ominus_{d i=1}^s p_i) \ominus_d (\ominus_{d i=1}^t q_i) \in P$ and $n = max\{n_1, n_2, \cdots, n_s, m_1, m_2, \cdots, m_t\}$. By Proposition 2.4 and Proposition 2.5, we get that

$$x \leq \ominus_{b i=1}^s (p_i \ominus_b n_{ib}(u \ominus_b \tau(u))) \wedge \ominus_{c i=1}^t (q_i \ominus_c m_{ic}(v \ominus_c \tau(v)))$$
$$\leq \ominus_{d i=1}^s (p_i \ominus_d n_{id}(u \ominus_d \tau(u))) \wedge \ominus_{d i=1}^t (q_i \ominus_d m_{id}(v \ominus_d \tau(v)))$$
$$\leq s_d(r \ominus_d n_d(u \ominus_d \tau(u))) \wedge t_d(r \ominus_d n_d(v \ominus_d \tau(v)))$$
$$\leq (st)_d((r \ominus_d n_d(u \ominus_d \tau(u))) \wedge (r \ominus_d n_d(v \ominus_d \tau(v))))$$
$$= (st)_d(r \ominus_d (n_d(u \ominus_d \tau(u)) \wedge n_d(v \ominus_d \tau(v))))$$
$$\leq (st)_d(r \ominus_d n^2{}_d((u \ominus_d \tau(u)) \wedge (v \ominus_d \tau(v)))).$$

It follows from Proposition 4.15 that $(u \ominus_d \tau(u)) \wedge (v \ominus_d \tau(v)) \in P$. This together with $r \in P$ implies $x \in P$, which is a contradiction. Therefore, $P \in PSI(A)$. $\qquad\square$

Let (A, τ) be a state Ehoop and X a subset of A. Set $\mathbf{O}(X) = \{P \in PSI(A) | X \nsubseteq P\}$.

Proposition 4.19. *Let (A, τ) be a state Ehoop, $X, Y \subseteq A$ and $\{X_i\}_{i \in I}$ a family subsets of A. Then*

(i) *$X \subseteq Y$ implies $\mathbf{O}(X) \subseteq \mathbf{O}(Y)$;*
(ii) *$\mathbf{O}(\emptyset) = \mathbf{O}(\{0\}) = \emptyset$, $\mathbf{O}(Id(A)) = PSI(A)$;*
(iii) *if A satisfies the double negation property, $\mathbf{O}(X) \cap \mathbf{O}(Y) = \mathbf{O}(< X >_\tau \cap < Y >_\tau)$;*

(iv) $\cup_{i \in I} \mathbf{O}(X_i) = \mathbf{O}(\cup_{i \in I} X_i)$;

(v) $\mathbf{O}(X) = \mathbf{O}(< X >_\tau)$;

(vi) *if A satisfies the double negation property, $\mathbf{O}(X) = \mathbf{O}(Y)$ iff $< X >_\tau = < Y >_\tau$.*

Proof. (i) If $I \in \mathbf{O}(X)$, then $X \nsubseteq I$. This together with $X \subseteq Y$ implies $Y \nsubseteq I$, i.e. $I \in \mathbf{O}(Y)$. Thus, $\mathbf{O}(X) \subseteq \mathbf{O}(Y)$.

(ii) It is obvious that $\mathbf{O}(\emptyset) = \mathbf{O}(\{0\}) = \emptyset$. Let $P \in PSI(A)$. By $P \neq A$, there is $x \in A \backslash P$. Since A has enough idempotent elements, there exists $a \in Id(A)$ such that $x \leq a$. It follows that $a \notin P$ and so $Id(A) \nsubseteq P$, i.e. $P \in \mathbf{O}(Id(A))$, which means $PSI(A) \subseteq \mathbf{O}(Id(A))$. Clearly, we have $\mathbf{O}(Id(A)) \subseteq PSI(A)$. Hence, $\mathbf{O}(Id(A)) = PSI(A)$.

(iii) Suppose $I \in \mathbf{O}(X) \cap \mathbf{O}(Y)$. Then $X, Y \nsubseteq I$ and so $< X >_\tau, < Y >_\tau \nsubseteq I$. Applying Proposition 4.17, we get $< X >_\tau \cap < Y >_\tau \nsubseteq I$. Thus, $I \in \mathbf{O}(< X >_\tau \cap < Y >_\tau)$, which means $\mathbf{O}(X) \cap \mathbf{O}(Y) \subseteq \mathbf{O}(< X >_\tau \cap < Y >_\tau)$. Conversely, if $I \in \mathbf{O}(< X >_\tau \cap < Y >_\tau)$, i.e. $< X >_\tau \cap < Y >_\tau \nsubseteq I$, we have $< X >_\tau \nsubseteq I$ and $< Y >_\tau \nsubseteq I$. It follows that $X, Y \nsubseteq I$ and so $I \in \mathbf{O}(X) \cap \mathbf{O}(Y)$. Therefore, $\mathbf{O}(< X >_\tau \cap < Y >_\tau) = \mathbf{O}(X) \cap \mathbf{O}(Y)$.

(iv) Since $X_i \subseteq \cup_{i \in I} X_i$ for any $i \in I$, we have $\mathbf{O}(X_i) \subseteq \mathbf{O}(\cup_{i \in I} X_i)$ and so $\cup_{i \in I} \mathbf{O}(X_i) \subseteq \mathbf{O}(\cup_{i \in I} X_i)$. Conversely, if $I \in \mathbf{O}(\cup_{i \in I} X_i)$, then $\cup_{i \in I} X_i \nsubseteq I$. There is $i_0 \in I$ such that $X_{i_0} \nsubseteq I$. Hence, $I \in \mathbf{O}(X_{i_0}) \subseteq \cup_{i \in I} \mathbf{O}(X_i)$ and so $\mathbf{O}(\cup_{i \in I} X_i) \subseteq \cup_{i \in I} \mathbf{O}(X_i)$. Thus, $\cup_{i \in I} \mathbf{O}(X_i) = \mathbf{O}(\cup_{i \in I} X_i)$.

(v) Let $I \in PSI(A)$. Then $I \in \mathbf{O}(X)$ iff $X \nsubseteq I$ iff $< X >_\tau \nsubseteq I$ iff $I \in \mathbf{O}(< X >_\tau)$. Hence, $\mathbf{O}(X) = \mathbf{O}(< X >_\tau)$.

(vi) Suppose $\mathbf{O}(X) = \mathbf{O}(Y)$. If $< X >_\tau \neq < Y >_\tau$, without loss of generality, we can assume that there exists $u \in < Y >_\tau \backslash < X >_\tau$. By Prime state ideal theorem, there is $P \in PSI(A)$ such that $u \notin P$ and $< X >_\tau \subseteq P$. Then $< Y >_\tau \nsubseteq P$. It follows that $P \notin \mathbf{O}(< X >_\tau) = \mathbf{O}(X)$ and $P \in \mathbf{O}(< Y >_\tau) = \mathbf{O}(Y)$, which is a contradiction. Thus, $< X >_\tau = < Y >_\tau$.

Conversely, the proof is straightforward by (v). $\qquad\square$

Proposition 4.20. *Let (A, τ) be a state Ehoop. The following properties hold:*

(i) *if $x, y \in A$ and $x \leq y$, then $\mathbf{O}(\{x\}) \subseteq \mathbf{O}(\{y\})$;*

(ii) *for any $x, y \in A$, any $a \in Id(A)$ with $x, y \leq a$, $\mathbf{O}(\{x\}) \cup \mathbf{O}(\{y\}) = \mathbf{O}(\{x \ominus_a y\})$;*

(iii) *for any $x \in A$, any $a \in Id(A)$ with $x, \tau(x) \leq a$, $\mathbf{O}(\{x\}) = \mathbf{O}(\{x \ominus_a \tau(x)\})$.*

Proof. (i) Let $I \in \mathbf{O}(\{x\})$. That is, $x \notin I$. As $x \leq y$, we have $y \notin I$. Thus, $I \in \mathbf{O}(\{y\})$ and so $\mathbf{O}(\{x\}) \subseteq \mathbf{O}(\{y\})$.

(ii) Let $I \in SI(A)$ and $x, y \in A$. We claim that for any $a \in Id(A)$ with $x, y \leq a$, $x \notin I$ or $y \notin I$ iff $x \ominus_a y \notin I$. In fact, suppose $x \notin I$ or $y \notin I$. It follows from

$x, y \leq x \ominus_a y$ that $x \ominus_a y \notin I$. Conversely, suppose $x \ominus_a y \notin I$. If $x, y \in I$, we have $x \ominus_a y \in I$ by $I \in SI(A)$, which is a contradiction. This means $x \notin I$ or $y \notin I$.

Let $P \in PSI(A)$. Then $P \in \mathbf{O}(\{x\}) \cup \mathbf{O}(\{y\})$ iff $x \notin P$ or $y \notin P$ iff $x \ominus_a y \notin P$ for any $a \in Id(A)$ such that $x, y \leq a$ iff $P \in \mathbf{O}(\{x \ominus_a y\})$ for any $a \in Id(A)$ such that $x, y \leq a$. Hence, $\mathbf{O}(\{x\}) \cup \mathbf{O}(\{y\}) = \mathbf{O}(\{x \ominus_a y\})$.

(iii) By Proposition 4.10 and Proposition 4.19(v), $\mathbf{O}(\{x \ominus_a \tau(x)\}) = \mathbf{O}(< x \ominus_a \tau(x) >_\tau) = \mathbf{O}(< x >_\tau) = \mathbf{O}(\{x\})$. $\qquad\square$

Let (A, τ) be a state Ehoop. Define $\Gamma = \{\mathbf{O}(X) | X \subseteq A\}$. If A satisfies the double negation property, Γ is a topology on $PSI(A)$ by Proposition 4.19.

Theorem 4.21. *Let A be an Ehoop with the double negation property and τ an internal state on A. Then $(PSI(A), \Gamma)$ is a compact T_0-space.*

Proof. We claim that $\{\mathbf{O}(\{x\}) | x \in A\}$ is a basis for Γ. Indeed, for any subset X of A, we obtain $\mathbf{O}(X) = \mathbf{O}(\cup_{u \in X}\{u\}) = \cup_{u \in X}\mathbf{O}(\{u\})$ by Proposition 4.19(iv), which means that any open set in Γ is the union of some elements of $\{\mathbf{O}(\{x\}) | x \in A\}$.

Suppose $x \in A$ and $\mathbf{O}(\{x\}) = \cup_{i \in I}\mathbf{O}(\{x_i\}) = \mathbf{O}(\cup_{i \in I}\{x_i\})$, where $x_i \in A$ for any $i \in I$. Due to Proposition 4.19(vi), $< x >_\tau =< \cup_{i \in I}\{x_i\} >_\tau$ and so $x \in< \cup_{i \in I}\{x_i\} >_\tau$. By Proposition 4.11(ii), we have $x \leq x_{i_1} \ominus_a \tau(x_{i_1}) \ominus_a x_{i_2} \ominus_a \tau(x_{i_2}) \ominus_a \cdots \ominus_a x_{i_n} \ominus_a \tau(x_{i_n})$ for some $n \in \mathbb{N}\setminus\{0\}$, $i_1, i_2, \cdots, i_n \in I$ and $a \in Id(A)$ such that $x_{i_1}, \tau(x_{i_1}), x_{i_2}, \tau(x_{i_2}), \cdots, x_{i_n}, \tau(x_{i_n}) \leq a$. If follows from Proposition 4.20 that

$$\begin{aligned}\mathbf{O}(\{x\}) &\subseteq \mathbf{O}(\{x_{i_1} \ominus_a \tau(x_{i_1}) \ominus_a x_{i_2} \ominus_a \tau(x_{i_2}) \ominus_a \cdots \ominus_a x_{i_n} \ominus_a \tau(x_{i_n})\}) \\ &= \mathbf{O}(\{x_{i_1} \ominus_a \tau(x_{i_1})\}) \cup \mathbf{O}(\{x_{i_2} \ominus_a \tau(x_{i_2})\}) \cup \cdots \cup \mathbf{O}(\{x_{i_n} \ominus_a \tau(x_{i_n})\}) \\ &= \mathbf{O}(\{x_{i_1}\}) \cup \mathbf{O}(\{x_{i_2}\}) \cdots \cup \mathbf{O}(\{x_{i_n}\}).\end{aligned}$$

This together with $\mathbf{O}(\{x_{i_1}\}) \cup \mathbf{O}(\{x_{i_2}\}) \cdots \cup \mathbf{O}(\{x_{i_n}\}) \subseteq \cup_{i \in I}\mathbf{O}(\{x_i\}) = \mathbf{O}(\{x\})$ implies that $\mathbf{O}(\{x\}) = \mathbf{O}(\{x_{i_1}\}) \cup \mathbf{O}(\{x_{i_2}\}) \cdots \cup \mathbf{O}(\{x_{i_n}\})$. Since $\{\mathbf{O}(\{x\}) | x \in A\}$ is a basis for Γ, Γ is compact.

We shall show that Γ is a T_0-space. Suppose $I_1, I_2 \in PSI(A)$ and $I_1 \neq I_2$. Without loss of generality, we can assume that there exists $x \in I_1 \setminus I_2$. Set $U = \mathbf{O}(\{x\})$. It follows that $I_1 \notin U$ and $I_2 \in U$. Thus, Γ is a T_0-space. $\qquad\square$

5　Conclusion

In this paper, we present Bosbach states and Riečan states on Ehoops with a bottom element. It is proved that these two kinds of states on Ehoops with a bottom element are the same thing. The notion of state-morphisms on Ehoops with a bottom element is also introduced. We investigate the relationship between state-morphisms and maximal filters of an Ehoop with a bottom element. In addition, we prove that each Ehoop with a bottom element admits at least one Bosbach/Riečan state. Moreover, we study internal states on Ehoops with a bottom element and give a prime state ideal theorem. When a state Ehoop (A, τ) satisfies the double negation property, we get a topological space using prime state ideals.

There are some topics for further study. (1) What is the relationship between state-morphisms and maximal ideals of an Ehoop with a bottom element? (2) Can we introduce the notions of state filters and prime state filters of an Ehoop with a bottom element, and derive a topological space using prime state filters? (3) Give more examples of Bosbach/Riečan states and internal states on an Ehoop with a bottom element. (4) We shall study the existence of nontrivial internal states of an Ehoop with a bottom element.

References

[1] M. Aaly Kologani, R. A. Borzooei, *On ideal theory of hoops,* Mathematica Bohemica, **145 (2)**, (2020), 141-162.

[2] W. J. Blok, D. Pigozzi, *On the structure of varieties with equationally definable principal congruences III,* Algebra Universalis, **32**, (1994), 545-608.

[3] R. A. Borzooei, M. Aaly Kologani, *Stabilizer topology of hoops,* Algebraic Structures and Their Applications, **1 (1)**, (2014), 35-48.

[4] B. Bosbach, *Komplementäre Halbgruppen. Axiomatik und Arithmetik,* Fundamenta Mathematicae, **64**, (1969), 257-287.

[5] B. Bosbach, *Komplementäre Halbgruppen. Kongruenzen und Quotienten,* Fundamenta Mathematicae, **69**, (1970), 1-14.

[6] M. Botur, A. Dvurečenskij, *On pseudo-BL-algebras and pseudo-hoops with normal maximal filters,* Soft Computing, **20 (2)**, (2016), 439–448.

[7] J. R. Büchi, T. M. Owens, *Complemented monoids and hoops,* unpublished manuscript.

[8] L. C. Ciungu, *Algebras on subintervals of pseudo-hoops,* Fuzzy Sets and Systems, **160**, (2009), 1099-1113.

[9] L. C. Ciungu, *Bounded pseudo-hoops with internal states,* Mathematica Slovaca, **63**, (2013), 903-934.

[10] L. C. Ciungu, *Non-commutative multiple-valued logic algebras,* Springer, New York, (2014).

[11] L. C. Ciungu, J. Kühr, *New probabilistic model for pseudo-BCK algebras and pseudo-hoops,* Journal of Multiple-Valued Logic and Soft Computing, **20**, (2013), 373-400.

[12] A. Dvurečenskij, *States on pseudo MV-algebras,* Studia Logica, **68**, (2001), 301–327.

[13] A. Dvurečenskij, R. Giuntini, T. Kowalski, *On the structure of pseudo BL-algebras and pseudo hoops in quantum logics,* Foundations of Physics, **40**, (2010), 1519-1542.

[14] A. Dvurečenskij, T. Kowalski, *On decomposition of pseudo BL-algebras,* Mathematica Slovaca, **61**, (2011), 307-326.

[15] A. Dvurečenskij, J. Rachůnek, *Probabilistic averaging in bounded Rℓ-monoids,* Semigroup Forum, **72**, (2006), 190-206.

[16] A. Dvurečenskij, O. Zahiri, *On EMV-algebras,* Fuzzy Sets and Systems, **373**, (2019), 116-148.

[17] A. Dvurečenskij, O. Zahiri, *States on EMV-algebras,* Soft Computing, **23**, (2019), 7513-7536.

[18] I. M. A. Ferreirim, *On varieties and quasivarieties of hoops and their reducts,* Ph.D.thesis, University of Illinois at Chicago, (1992).

[19] G. Georgescu, L. Leuştean, V. Preoteasa, *Pseudo-hoops,* Journal of Multiple-Valued Logic and Soft Computing, **11**, (2005), 153-184.

[20] P. He, B. Zhao, X. Xin, *States and internal states on semihoops,* Soft Computing, **21**, (2017), 2941-2957.

[21] M. Kondo, *Some types of filters in hoops,* 2011 41st IEEE International Symposium on Multiple-Valued Logic, (2011), 50-53.

[22] H. Liu, *EBL-algebras,* Soft Computing, **24**, (2020), 14333-14343.

[23] D. Mundici, *Averaging the truth-value in Łukasiewicz logic,* Studia Logica, **55**, (1995), 113-127.

[24] A. Namdar, R. A. Borzooei, *Nodal filters in hoop algebras,* Soft Computing, **22**, (2018), 7119-7128.

[25] A. Namdar, R. A. Borzooei, A. Borumand Saeid, M. Aaly Kologani, *Some results in hoop algebras,* Journal of Intelligent & Fuzzy Systems, **32**, (2017), 1805-1813.

[26] B. Riečan, *On the probability on BL-algebras,* Acta Math. Nitra, **4**, (2000), 3-13.

[27] F. Xie, H. Liu, *Ehoops,* Journal of Multiple-Valued Logic and Soft Computing, **37**, (2021), 77-106.

 Received 6 September 2021

QUANTUM IMMORTALITY AND NON-CLASSICAL LOGIC

PHILLIP L. WILSON

School of Mathematics & Statistics, University of Canterbury, New Zealand.
Te Pūnaha Matatini, New Zealand.
phillip.wilson@canterbury.ac.nz.

Abstract

The *Everett Box* is a device in which an observer and a lethal quantum apparatus are isolated from the rest of the universe. On a regular basis, successive *trials* occur, in each of which an automatic measurement of a quantum superposition inside the apparatus either causes instant death or does nothing to the observer. From the observer's perspective, the chances of surviving m trials monotonically decreases with increasing m. As a result, if the observer is still alive for sufficiently large m she rejects any interpretation of quantum mechanics which is not the many-worlds interpretation (MWI), since surviving m trials becomes vanishingly unlikely in a single world, whereas a version of her will necessarily survive in the branching MWI universe. That is, the MWI is testable, at least privately. Here we ask whether this conclusion still holds if rather than a classical understanding of limits built on classical logic we instead require our physics to satisfy a computability requirement by investigating the Everett Box in a model of a computational universe using a variety of constructive logic, Recursive Constructive Mathematics. We show that although the standard argument sketched above is no longer valid, we nevertheless can argue that the MWI remains privately testable in a computable universe.

1 Introduction

The famously accurate predictions obtained by calculating solutions to the Schrödinger equation do not depend upon the interpretation of quantum mechanics. For a physicist who wants to distinguish between these interpretations, being told to "shut up and calculate" misses the point: she is interested in ontology (what is

---

The author wishes to thank the referees of an earlier version of this paper for their insightful comments, the responses to which have certainly improved the paper.

real) not epistemology (what we can know). But since the predictions of all interpretations are the same, given by an interpretion-free solution of the Schrödinger equation, how can we distinguish between them experimentally? In particular, is the *many-worlds interpretation*[1] (MWI) of Hugh Everett [18, 17] distinguishable from other interpretations? More fundamentally, is it testable?

These questions were addressed by Tegmark in [33] (see also references therein for more history of the questions). In particular, Tegmark argued that the MWI is testable, and can be distinguished from other interpretations under extreme conditions, at least for one observer. Tegmark's argument (see also [34]) has two main points:

T1 the MWI is testable and can be distinguished from other interpretations through experiment;

T2 the same experiment which can distinguish the MWI from other interpretations should also be taken as evidence in support of the MWI *for the experimenter herself*.[2]

Popular descriptions of the MWI talk of the universe splitting into n branches (or *worlds*) whenever a quantum measurement has n possible outcomes. However, as emphasised by [33], this was never postulated by Everett himself. Indeed, the core of the MWI is simply that the wavefunction never collapses and so the universe can be taken to be governed by a single, objectively real, universal wave function. However, the language of a splitting or branching universe, or observer, is a useful tool, and since it is also a common one (see a brief discussion of its history in [22]) we will sometimes use that language herein.

One of the better-known thought experiments which claims to establish T1 and T2 is a variant of the classic Schrödinger's cat experiment in which the cat, or rather a human in place of the cat, is the observer [34, 33, 22]. The life of the observer depends upon the outcome of an automatic measurement of a qubit called a *trial*. Many trials occur, one after the other on a regular basis. We describe in detail in §2 how this *Everett Box*[3] is able to establish T1 and T2. The argument is a probabilistic one: though the observer might get lucky and survive a few trials, continued survival in a non-MWI universe is extremely unlikely. On the other hand, a version of the observer is guaranteed to survive with probability 1 in one branch of an MWI universe, guaranteeing the so-called *Quantum Immortality* of that observer.

---

[1]So named by Bryce DeWitt [15, 16].

[2]In the language of [22], the MWI is *privately* testable, if not *publicly* testable.

[3]Named in homage to the genesis of these ideas in Everett's seminal thesis [18] even though Everett himself never formulated them.

From the private perspective of the observer, her continuing survival therefore counts as evidence in support of the MWI, establishing both T2 and T1, non-MWI options being rejected as being too unlikely.

We call the line of argument summarised above and presented in §2 the *Quantum Immortality Argument (QIA)*, and refer to its conclusion as *Quantum Immortality*. Although Everett never formally defined this experiment, variants of it have been given independently by several authors [31, 33]. Neither the argument nor its conclusion is universally accepted. The QIA has been attacked from various directions, not least in terms of its real-world applicability, for instance around the definition of death (or at least of a discrete binary distinction between "alive" and "dead") — see [34, 33]. From the philosophical perspective we see critiques[4] based on the classical philosophical problem of individual identity and its persistence, what it means to "expect" a subjective outcome like one's own death as opposed to predicting an objective event, the distinction between actual and probable events, and the meaning of probabilistic thinking in the MWI context[5]; see [14, 22, 27, 1, 30, 35], and references therein. Everett himself anticipated some of these objections in [18].

There are two other ways in which the QIA is critiqued, both of which are much more general in their scope. They concern (1) the role of infinity and the infinitesimal in physics, and (2) the role of the computable. A motivation for the first of these is that if we live in a finite universe which has existed for finite time, and if the fields, matter, time, and space of the universe are all discrete at sufficiently small scales, then we should reject all objects and arguments which employ the infinite and the infinitesimal. Such strictly *finitist* theories include digital physics [36], cellular automata [38], loop quantum gravity [28], and more besides — see [29] and references therein. The questionable role of infinity in the QIA has been highlighted by [34] amongst others.

The second critique, namely that our current theories of physics are non-computable, is the focus of the present work. Requiring a computable theory of physics is essentially the same as requiring all knowledge to be obtained through an algorithmic process in finite time. It is not the same as requiring only finite objects, but it does necessitate working within so-called non-classical logics, as we outline in detail below. The desirable quality of computability in the foundations of physics is not obtained by classical logic.

Thus while probabilities and probabilistic thinking have been highlighted as potential concerns with the QIA [22, 27], and while the role of the infinite and the

---

[4]Not all of the authors of these critiques reject the MWI even while rejecting the QIA; see for example [14].

[5]Of particular note here is the lack of an ensemble or "God's eye" view of the experiment: there is no vantage point from which an external observer can quantify overall outcomes.

infinitesimal in physics have also been called into question in this context [34], to our knowledge no-one has examined the argument from a computable perspective before, and in particular from within non-classical logic.

Here we show that the testability of the MWI and the subjective evidence in support of the MWI given by the QIA are based on a classical understanding of limiting behaviours of functions which need not hold in other, non-classical logics. In particular, we show that the Quantum Immortality Argument fails in a constructive logic called Recursive Constructive Mathematics, commonly referred to as RUSS, in which all results are computable. Within RUSS, we show that the existence of so-called *pathological* probability distributions mean that we must reject the QIA. However, we are able to show through a new argument that a constructive version of the QIA holds even in a universe (or universes) governed by such non-classical logics, and thus that the MWI remains (privately) testable and distinguishable from other interpretations of quantum mechanics.

In §2 we give a brief overview of the QIA, and in particular how the Everett Box implies T1 and T2. Next, in §3 we define computability and outline the arguments in favour of requiring computability in theories of physics, before giving a summary of the main result from [24] on which we base the principal argument in this paper. With this background we prove in §4 the Pathological Mortality Theorem, which shows that the QIA does not work in RUSS. However, in §5 we present a constructive, computable proof that Quantum Immortality nevertheless holds in a universe whose logic is that of RUSS. We call this argument the *computable Quantum Immortality Argument*. It shows that the MWI is testable in a universe governed by computable logic. Finally, we summarise and discuss our results in §6.

2   Quantum Immortality

A conscious observer is placed in a box with a lethal quantum apparatus. The contents of the box are completely isolated from the rest of the universe. Although this thought experiment does not depend on the details of the lethal apparatus, a particularly clear example is given by [33] and called the "quantum gun". The quantum gun consists of a gun coupled to a quantum system of a particle in a superposition of two states. At regular time intervals, a measurement of this qubit is made automatically, and if it is found to be in one state the gun fires a bullet, while if it is in the other it does not fire. After either firing or not firing, the quantum gun resets: a new superposition is set up and the memoryless process repeats[6]. Each

---

[6]This slight variant of Tegmark's quantum gun of [33] was given in [34]. Note that both of Tegmark's versions include an assistant in the box with the observer, but this is not a necessary

independent occurrence of this process we call a *trial*. We take this or a similar lethal setup to be indefinitely repeatable and to occur every second[7]. The apparatus and the observer are isolated from the rest of the universe, and this setup constitutes the Everett Box.

What is the experience of the observer? It is rather starkly illustrated by Tegmark's gun if we contrast the Everett Box with a similar experiment in which instead of being aimed at the observer the gun merely fires or does not fire depending on the measurement, and is aimed at a target in the box while the observer observes the experiment from within the box. In this case, the observer can expect to hear a random string of bangs and clicks: the bangs correspond to the gun firing, the clicks to it not firing and the equipment resetting. Over time the relative proportion of bangs and clicks will tend towards the relative likelihoods of those two outcomes. In the standard formulation, both outcomes occur with equal probability and thus the observer expects over time that 50% of the sounds will be bangs, and 50% clicks. The QIA is actually independent of these likelihoods, which need not be either equal or constant [18, 17, 34]. It is such a general case that we consider in this paper.

The immediately preceding description is not that of the Everett Box, because the life and hence consciousness[8] of the observer does not depend upon the outcome of the measurement of the qubit. In the Everett Box, the gun is aimed at the observer in such a way that should it fire then death is certain and swift[9]. In this case, what should the observer expect[10]?

The answer depends upon which interpretation of quantum mechanics holds in our universe. If there is only one world, then for the majority of interpretations each trial involves the collapse of the wavefunction and a single outcome occurs for the observer: either she hears a "click" or she is instantly killed (and so hears nothing)[11]. She might get lucky once, she might get lucky twice, but as time goes on and the number of trials increases, the odds of her surviving decrease exponentially.

If, however, there are many worlds, the totality of which contain all possible outcomes and histories, then by necessity there is always an observer alive after any number of trials. For example, after one trial there are two versions of the

---

feature of the setup and has been omitted through automation here.

[7]The time interval is not important to the subsequent argument, other than to allow for many repetitions within a human lifetime.

[8]Consciousness surviving death is not a part of the thought experiment.

[9]Both conditions are necessary as outlined in [33, 22].

[10]Assuming that she has first verified that her apparatus is working by observing its operation without being in the line of fire.

[11]Note that even non-collapse interpretations such as Bohmian mechanics predict a single outcome for the observer.

observer, the universe[12] having branched into two at the moment of the quantum measurement. In one world the observer heard "click" while in the other she died. After two seconds there are four worlds. In one of them, the observer's history shows that she heard "click-click". In another world, she heard "click" and then died on trial 2. In the third and fourth she died on trial 1[13]. After three trials there is an observer whose history is "click-click-click", and after any number, m, of trials there will always be one world in which the observer has survived to hear m clicks. After a large number of trials there are many worlds, in all but one of which the observer is dead, but crucially there remains one living observer. Thus the subjective probability of surviving m trials is 1 for any m, because there is a world in which the observer is still alive after any number of trials [22]. Note that the 2^m worlds "created" in this branching process are not used for quantifying the *subjective* probability of 1 that the the observer survives — at no point do we calculate subjective probabilities over a set of universes in only some of which does a subject exist [22].

Thus from the observer's perspective[14] this experiment can establish T1 and T2, though the stakes are high. The argument runs as follows. The chances of remaining alive after a large number of trials in a non-MWI universe is monotonically and exponentially decreasing because each trial is independent of the preceding trials. Thus at some point the probability of being alive will be lower than some threshold at which the still-alive observer can reject any non-MWI interpretation purely on the grounds of the low probability of such a sequence of events occurring. This is a standard experimental approach, and as usual the threshold $\epsilon \ll 1$ could be set to the traditional 5σ-level, or indeed to a level of any stringency based on any criterion[15] due to the monotonically decreasing dependence on m of the probability of remaining alive. Furthermore, with each subsequent survived trial the confidence in rejecting any non-MWI interpretation increases. Of course, if we do not live in an MWI universe then the experiment simply kills the observer within a short time. She does not know that she does not live in an MWI universe, but neither does she know anything ever again.

In more rigorous terms, in non-MWI interpretations the probability of being dead after m trials, $P(m)$, in the standard presentation in which the probability of

---

[12]Or at least, the observer.

[13]Here we assume that a predetermined number of trials occur regardless of whether the observer is alive. This assumption is not actually necessary for the arguments which follow, and can easily (though with some loss of elegance) be removed.

[14]And from hers alone; an external observer opening the box after a pre-ordained number of seconds (trials) will almost certainly find a corpse within the box. This footnote is the only point in the present paper at which we quantify over a multiverse of branched universes.

[15]It is also possible to formulate a threshold based on a Bayesian analysis.

death at each trial is 50%, is simply

$$P(m) = 1 - \left(\frac{1}{2}\right)^m$$

which tends to unity as $m \to \infty$. The same conclusion holds regardless of the probability of staying alive on trial k, which we denote p_k. In this case, because $p_k < 1$ for all k we still have

$$P(m) = 1 - \prod_{k=1}^{m} p_k \to 1 \quad \text{as} \quad m \to \infty. \tag{1}$$

While (1) will remain true throughout this paper, we will see that in the computable logic RUSS we can no longer use it to conclude that the observer necessarily must expect to be dead after any finite number of trials. First, we must review what it means to be computable.

3 Computability and The Infinite Monkey Theorem

3.1 Computability and Logic

A problem is said to be *computable* if it can be solved in an effective manner, which can be more formally defined in a number of models of computation [9, 12, 4]. Loosely speaking, computable problems are those which can be solved algorithmically in finite time. The major milestone in computability theory is the Turing-Church thesis identifying computable functions on the natural numbers with functions computable on a Turing machine [4, 8, 13, 12].

It is not simply the rise in computer simulations, nor the "shut-up-and-calculate" instrumentalist approach to physics [25], which have led some authors to suggest that computability should be a requirement for our theories of physics [39, 28, 29, 32, 19, 23, 38]. It is instead the notion of the *effective method* embedded in computability that is important. A method is called effective for a class of problems when it comprises a finite set of instructions which can be followed by a mechanical device[16], that these instructions produce a correct answer, and that they finish after a finite number of steps [12].

From a philosophical perspective, the desirability of computability in physics is therefore a product of a desire to know, and a belief that the universe is ultimately comprehensible to us. The reasoning in the syllogism goes that if we accept the two

---

[16]The idea here is not that they must be followed by such a device, but that even a human following them needs no *ingenuity* in order to derive a correct answer.

premises that (1) the universe is entirely comprehensible to the human mind, and (2) there is nothing extra-computational happening in the human mind, then we must accept the conclusion that physics is necessarily computable.

Moreover, from a practical point of view, the quantum calculations often used to verify observations are based on the classical solution to systems of partial differential equations. Both the systems themselves, with their corresponding boundary and initial conditions, and the numerical soltuions, are formulated and numerically approximated using methods relying ultimately on classical logic — see for example the discussion of the differential operator in [5], and other issues outlined in [8].

However, our current theories of physics are not computable, built as they are on classical mathematical ideas which in turn rely on classical, non-computable, logic [8]. There are two issues here. The first concerns the notion of infinity and the related notion of continuity. Infinities abound in our physical theories, whether they are in limiting behaviours (as examined in non-classical logics in the present paper) or in the related idea of continuous matter or continuous fields. In the latter case, even though we seem to know that neither matter nor fields are continuous in our universe, we treat the "gap" between our continuous theories and discrete nature as being essentially a rounding error: the high accuracy of predictions made with the (presumptively Platonic) continuous theories is because our universe is approximately continuous. It is, after all, perhaps only discrete below the Planck length, or on time scales shorter than the Planck time.

The second issue, and the one that concerns us in this paper, is the notion of the underlying logic of the universe. Classical logic is not computable, relying as it does on non-computable notions such as the Law of Excluded Middle (LEM) and omniscience principles [7, 8]. Why should we work with a logic that does not allow for computability if we wish our physics to be computable? One answer is similar to the response to continuity and infinity: because this logic works, to an astonishing degree [37]. A second response is simply to reject the second premise given above. Perhaps there is something extra-computational happening within the human mind[17]. This is consistent with a robustly Platonic vision of the universe. If mathematical objects exist in a Platonic realm of forms to which our minds (somewhat mysteriously) have access, then the necessity of computability can be rejected. This is also consistent with the view above that our physical universe is only an (albeit excellent) approximation to one of Platonic forms.

If however we insist with the authors above that our logic must be computable, then we necessarily have to work with non-classical logics which are computable. In particular, we should work within so-called *constructive* interpretations of logic

---

[17]This perspective overlaps somewhat with the notion of *hypercomputation* [10, 11]

[8], in which the classical interpretations of disjunction and existence are rejected in favour of constructive ones. For example, the quantifier "there exists" becomes "we can construct (that is, give an effective method for defining) an object for which the given statement is true". There are several varieties of constructive mathematics [7]. It should be noted that not all varieties reject notions of infinity. Bishop's Constructive Mathematics (referred to as BISH) [2], for example, admits many classical mathematical objects which rely on infinities and continuity, but insists that proofs using these objects must proceed constructively (and are therefore computable). This illustrates the important distinction between the *epistemological constructivism* of BISH which remains agnostic on the ontology of mathematical objects, and the *ontological constructivism* of other varieties of constructive mathematics which insist that both objects and proofs (procedures) must be computable [8, 7]. It has been said that computable mathematics is simply mathematics done with intuitionistic logic [8].

In order to subject the QIA to a strong scrutiny in a non-classical logic, we here choose an ontologically constructive variety of constructive logic, Recursive Constructive Mathematics, RUSS [8]. RUSS is a constructive version of recursive function theory, in which functions on the natural numbers are defined recursively. Essentially, RUSS takes the classical recursive analysis in the tradition of Turing and Church but uses only intuitionistic logic.

We remind the reader that this paper is not an argument in support of constructive methods in physics in general, neither of RUSS in particular. (The interested reader is referred to [20, 26, 5, 6, 3] amongst others.) Rather, it seeks to examine the QIA in a computable context, in anticipation either of reaching a testable implication that the universe is computable, or of recasting the QIA in a computable form. In the following subsection, we briefly outline the theorems of a recent work in computable probability based on RUSS which will be central to the argument of this paper.

3.2 The Infinite Monkey Theorem

Working in RUSS, [24] proved a seemingly counter-intuitive theorem, which we call here the *Infinite Monkey Theorem (IMT)*. To state the IMT we first need some notation. The IMT was written in the playful language of the famous aphorism that a large enough group of monkeys with typewriters will reproduce the complete works of Shakespeare, but as is made clear in [24], the IMT is really about computable probability distributions, as indeed is our focus in the present paper.

Retaining the metaphor of [24], we work in an alphabet A (of size $|A|$, including punctuation) and call a *w-string* any string of characters of length $w \in \mathbb{N}$. For example, "*Everett*" is a 7-string over the alphabet $\{E, e, r, t, v\}$. Each monkey works on a computer keyboard with $|A|$ unique keys and each monkey types a w-string in finite time. We define M to be an infinite, enumerable set of monkeys (the *monkeyverse*), and for any $m \in \mathbb{N}$ the m-troop of monkeys to be the first m monkeys in M. We then have

Theorem 1 (Infinite Monkey Theorem). *Given a finite target w-string T_w and a positive real number ϵ, there exists a computable probability distribution on M of producing w-strings such that:*

(i) *the classical probability that no monkey in M produces T_w is 0; and*

(ii) *the probability of a monkey in* any *m-troop producing T_w is less than ϵ.*

[24] established an even stronger, target-free version of this theorem, which requires only a knowledge of w, not of T_w.

The theorem and its proof are computable. The theorem shows that while it is classically true that it is impossible that no monkey reproduces the works of Shakespeare (part (i)), it is possible to construct a so-called *pathological* probability distribution on the monkeyverse such that the chances of actually finding the monkey that does so can be made arbitrarily small (part (ii)). The key point in part (ii) is that this is true for *any* finite m-troop of monkeys; the pathological distribution does not require knowledge of the size of the m-troop, it is simply pathological for all finite sets.

The monkeys correspond to any finite black-box process occurring in finite time. The general conclusion drawn in [24] is that in a computable universe the space of all possible probability distributions on enumerable sets contains a non-empty set of pathological distributions for which the IMT holds. This is in contradistinction to a universe governed by classical logic in which the IMT does not hold. It is this distinction that we exploit in the remainder of the paper, by examining the impact of the existence of pathological distributions on the enumerable set of trials in the Everett Box.

4   Pathological Distributions Imply the Rejection of the Quantum Immortality Argument

With the notation from §2 we can state that the probability $P(m)$ of dying within m trials is given by

$$P(m) = 1 - \prod_{k=1}^{m} p_k \tag{2}$$

for any $m \in \mathbb{N}$, where p_k is the probability of not dying on trial k. Note that here, in contrast to the manner in which the Everett Box is normally described, but in keeping with the more general case which Everett himself allowed for in [18], we consider a quantum apparatus with variable probabilities at each trial. The subsequent argument does not depend upon the unknowability in advance of the probability of death at each trial, $1 - p_k$; after all, in the standard formulation, $p_k = 0.5$ for all k. Whatever the distribution of values of p_k, the observer cannot predict in advance whether she lives or dies on trial k, and her fate is determined purely by the unknowable quantum state of the apparatus. We can now state the following theorem.

Theorem 2 (Pathological Mortality Theorem). *The QIA fails in RUSS.*

Proof. The QIA relies upon the observer rejecting non-MWI interpretations once $1 - P(m)$, the probability of her surviving m trials, drops below some threshold. She can always bound the number of trials required to drop below this threshold even if she is ignorant of the distribution of p_k.

However, while classically $1 - P(m) \to 0$ as the number of trials tends to infinity, there is a computable probability distribution on the trials such that the probability that the observer is alive after *any* finite number of trials is arbitrarily close to 1. This follows directly from the proof of the IMT in [24]. In particular, we place the objects of the IMT and the objects of the Pathological Mortality Theorem (PMT) in one-to-one correspondence as outlined in the following table.

	IMT	PMT
p_k	probability that k^{th} monkey fails to reproduce Shakespeare	probability of not dying on k^{th} trial
$P(m)$	probability that m-troop does reproduce Shakespeare	probability of dying within m trials

Thus while classically it remains true that the observer's probability of being alive after m trials tends to 0 as m tends to infinity, the classical interpretation of that result as being that after a certain *finite* number of trials the probability of the observer being alive should be so small that she should be surprised at remaining alive and reject any non-MWI interpretation is not true in a computational sense, in which that probability can remain arbitrarily close to 1 for *any* finite number of trials. □

There is an apparent contradiction between the classical probability of remaining alive tending to zero while it remains arbitrarily close to unity for *any* finite number of rials. However, as outlined in [24], it is important to note that the apparent contradiction here is only between the classical notion of the limit and the existence of computable pathological distributions within RUSS; we are deliberately comparing results from non-commensurate logical systems in order to show that classical logic may lead us astray in a computable universe.

The PMT says that in a universe run on computable logic[18] the QIA is no longer valid, and therefore if our universe is one governed by computable logic the sole argument which claims to show that MWI is testable is no longer valid. If the quantum apparatus happens to be governed by a pathological distribution then it is no longer unlikely that the observer remains alive after any finite number of trials, since that likelihood can remain arbitrarily close to 1. As a result, since the observer can never know for sure that she is not in a pathological distribution, she cannot surely state that remaining alive after any finite number of trials is unlikely and so she can never reject non-MWI interpretations of quantum mechanics. However, in the next section we formulate a new version of the QIA which hold even in a computable universe and even with the existence of the PMT.

5   Quantum Immortality Restored

We state our main result as a theorem.

Theorem 3 (computable Quantum Immortality Argument, cQIA). *The Everett Box implies that the MWI is testable and provides private evidence in support of the MWI even in a computable universe modelled by RUSS.*

Proof. Suppose an observer in an Everett Box in a RUSS-universe has survived many trials. There are two situations to consider depending on whether the probability distribution on the trials is pathological or not. To reiterate, the observer does not know and has no way of knowing which situation holds.

First, if the probability distribution is not pathological then the standard, classical-logic QIA holds, and the observer concludes that she must reject all non-MWI interpretations of quantum mechanics.

On the other hand, suppose that the distribution is pathological. Since Theorem 5 of [24] showed that such distributions are vanishingly rare then the observer must reject all non-MWI interpretations since in a single world the odds of being in such a distribution are vanishingly small, whereas in the MWI there will always be a branch of the observer alive in a pathological distribution. □

---

[18]We take this to be equivalent to the requirement for computability in the logic we use in our theories of physics.

To state the proof in other terms, we note that in a classical universe, so the original QIA goes, the observer rejects non-MWI interpretations because the odds of surviving repeated trials are so low, whereas in a RUSS computable universe she rejects non-MWI interpretations for the same reason if she happens to be in a non-pathological distribution, or because the odds of being in a pathological situation where the PMT holds in a single world are also vanishingly low. The observer does not need, therefore, any knowledge of whether she is in such a pathological experiment since either way she must reject non-MWI interpretations on the same basis: namely, the unlikelihood of being in that situation if there is only one world.

6 Discussion

Before a broader discussion, we once again reiterate that at no point does the observer need to quantify probabilities over other worlds. She merely compares her probability of surviving m trials *given there is one world* with a threshold set by some standard criterion such as the common 5σ-significance. In the cQIA, she makes two comparisons. In the first, she compares to the threshold the likelihood of there being only one world and in that world her equipment is governed by a non-pathological distribution. In the second, she compares to the threshold the likelihood of there being only one world and that in her world her equipment is governed by a pathological distribution. In both cases, she can easily pass any threshold by running a sufficient number of trials, and so she rejects the null hypotheses, thereby rejecting the existence of a single world. This she takes as private evidence for the MWI, which therefore remains testable even in a computable universe.

Therefore, we have argued that the Everett box is testable and provides evidence in support of the MWI interpretation of quantum mechanics even when the computability requirement is added to physics through employing RUSS, a constructive, computable logic. Our main point is therefore that those whose rejection of the MWI depends on some future recasting of physics in a computable form do not have that option if RUSS is the correct logic on which to base physics. We have shown, in fact, that a computable version of the QIA which we call the cQIA still holds in at least one computable logic. The case to be made against the MWI therefore must be stronger than has previously been appreciated.

This naturally raises the question as to whether a similar argument holds in other computable logics[19]. For example, can we reproduce the argument in BISH, a computable logic which preserves most classical mathematical objects, including some of those which involve either infinity or continuity? What about in other logics

---

[19]In which case, perhaps the "c" in cQIA would deserve capitalization.

which do not allow for such objects? And of course, what happens in a completely finitist universe?

We have two final points to make. The first is to point out that although the argument here is given in a quantum context, the argument behind the proof of the PMT works in any situation, quantum or otherwise, in which the probability of an event occurring tending to 1 in the limit of an infinite sequence of trials is taken to mean that the probability of that event not having happened in any finite sequence of trials necessarily tends to 0. In RUSS, this is not true.

Finally, we remind the reader that the MWI does not assume that new universes are "created" at each quantum decision — it takes the language of "branching" as a useful metaphorical tool rather than a literal description of reality, a reality simply described by a wavefunction which never collapses (see [33] and references therein). In fact, Tegmark has argued ([33, 34] and elsewhere) that a non-collapsing wavefunction plus decoherence strongly implies a platonic ontology in which the wavefunction and other mathematical objects are truly real, in fact constitute the only real things, and that the categories of human perceptions of the world are not to be taken literally, since they are prejudiced by our evolutionary history (see also [21]). This "Mathematical Universe Hypothesis" is an extreme form of platonism. We note in closing that despite popular arguments to the contrary, a platonic ontology is entirely compatible with a constructive epistemology — see [8].

References

[1] I. Aranyosi. Should we fear quantum torment. *Ratio*, 25(3):249–259, 2012.

[2] E. Bishop and D. S. Bridges. *Constructive Analysis*. A Series of Comprehensive Studies in Mathematics. Springer-Verlag, New York, 1985.

[3] D. Bridges and K. Svozil. Constructive mathematics and quantum physics. *International Journal of Theoretical Physics*, 39:503–515, 2000.

[4] D. S. Bridges. *Computability: a Mathematical Sketchbook*. Springer-Verlag, 1994.

[5] D. S. Bridges. Can constructive mathematics be applied in physics? *Journal of Philosophical Logic*, 28(5):439–453, 1997.

[6] D. S. Bridges. Constructive mathematics: a foundation for computable analysis. *Theoretical Computer Science*, 219(1–2):95–109, 1999.

[7] D. S. Bridges and F. Richman. *Varieties of Constructive Mathematics*. LMS Lecture Notes Series. Cambridge University Press, Cambridge, 1987.

[8] Douglas Bridges and Erik Palmgren. Constructive mathematics. In Edward N. Zalta, editor, *The Stanford Encyclopedia of Philosophy*. Metaphysics Research Lab, Stanford University, summer 2018 edition, 2018.

[9] S. B. Cooper. *Computability Theory*. Chapman & Hall, 2004.

[10] B. J. Copeland. Hypercomputation. *Minds and Machines*, 12:461–502, 2002.

[11] B. J. Copeland. Hypercomputation: philosophical issues. *Theoretical Computer Science*, 317:251–267, 2004.

[12] B. J. Copeland, C. Posy, and O. Shagrir. *Computability: Turing, Godel, Church, and Beyond*. MIT Press, 2013.

[13] Walter Dean. Recursive functions. In Edward N. Zalta, editor, *The Stanford Encyclopedia of Philosophy*. Metaphysics Research Lab, Stanford University, summer 2020 edition, 2020.

[14] D. Deutsch. Quantum theory of probability and decisions. *Proceedings of the Royal Society of London A*, 455:3129–3137, 1999.

[15] B. DeWitt. Quantum mechanics and reality. *Physics Today*, 23:30–40, 1970.

[16] B. DeWitt. The many-universe interpretation of quantum mechanics. In B. d'Espangat, editor, *Foundations of Quantum Mechanics*. Academic Press, New York, 1972.

[17] H. Everett, III. "Relative state" formulation of quantum mechanics. *Reviews of Modern Physics*, 29(3):454–462, 1957.

[18] H. Everett, III. *Theory of the universal wavefunction*. PhD thesis, Princeton University, Princeton, NJ, 1957.

[19] E. Fredkin. An introduction to digital philosophy. *International Journal of Theoretical Physics*, 42(2):189–247, 2003.

[20] G. Hellman. Constructive mathematics and quantum mechanics: Unbounded operators and the spectral theorem. *Journal of Philosophical Logic*, 22(3):221–248, 1993.

[21] D. Hoffman. *The Case Against Reality*. Penguin Random House, London, 2019.

[22] P. J. Lewis. What is it like to be Schrodinger's cat? *Analysis*, 60(1):22–29, 2000.

[23] S. Lloyd. A theory of quantum gravity based on quantum computation. *arXiv:quant-ph/0501135*, 2005.

[24] M. McKubre-Jordens and P. L. Wilson. Infinity in computable probability. *Journal of Applied Logics — IfCoLog Journal of Logics and their Applications*, 6(7):1253–1261, 2019.

[25] N. D. Mermin. Could Feynman have said this? *Physics Today*, 57(5):10–11, 2004.

[26] M. A. Nielsen. Computable functions, quantum measurements, and quantum dynamics. *Physical Review Letters*, 79(15):2915, 1997.

[27] D. Papineau. David Lewis and Schrodinger's cat. *Australian Journal of Philosophy*, 82:153–169, 2004.

[28] C. Rovelli and L. Smolin. Knot theory and quantum gravity. *Physical Review Letters*, 61(10):1155–1158, 1988.

[29] J. Schmidhuber. A computer scientist's view of life, the universe, and everything. In C. Freska, editor, *Foundations of Computer Science*. Springer, 1997.

[30] C. T. Sebens. Killer collapse: empirically probing the philosophically unsatisfactory region of grw. *Synthese*, 192:2599–2615, 2015.

[31] E. J. Squires. *The Mystery of the Quantum World*. Hilger, 1986.

[32] G. 't Hooft. Quantum gravity as a dissipative deterministic system. *Classical and Quantum Gravity*, 16(10):3263–3279, 1999.

[33] M. Tegmark. The interpretation of quantum mechanics: Many worlds or many words? *Fortsch.Phys.*, 46:855–862, 1998.

[34] M. Tegmark. *Our Mathematical Universe*. Vintage Books, New York, 2014.

[35] Lev Vaidman. Many-worlds interpretation of quantum mechanics. In Edward N. Zalta, editor, *The Stanford Encyclopedia of Philosophy*. Metaphysics Research Lab, Stanford University, fall 2018 edition, 2018.

[36] J. A. Wheeler. Information, physics, quantum: the search for links. In W. H. Zurek, editor, *Complexity, Entropy, and the Physics of Information*. Addison-Wesley, 1990.

[37] P. L. Wilson. What the applicability of mathematics says about its philosophy. In S. O. Hansson, editor, *Technology and Mathematics*. Springer, 2018.

[38] S. Wolfram. *A New Kind of Science*. Wolfram Media, 2002.

[39] K. Zuse. *Rechnender Raum*. Schriften zur Datenverarbeitung. Vieweg & Sohn, (English translation: Calculating Space, MIT project MAC AZT-70-164-GEMIT 1970), 1969.

Received 21 October 2021

www.ingramcontent.com/pod-product-compliance
Lightning Source LLC
Chambersburg PA
CBHW081237090426
42738CB00016B/3337